Studies in Agricultural Geography

K. E. Wheeler
A. M. Ladley
F. G. Leong

Studies in Agricultural Geography

Blond Educational

First published in Great Britain 1970 by
Blond Educational Limited
56 Doughty Street, London WC1
First published 1970 by
F. W. Cheshire Publishing Pty Ltd
Australia

Printed at The Griffin Press, Adelaide, South Australia
SBN 219 51773 8

Acknowledgements

Authors and publishers would like to make grateful acknowledgement to the following for assistance and permission to use copyright material:
Daphne Howie for preparation of maps used throughout the text; Mr Don Dobie, MHR, for assistance with information related to the Territories of Papua and New Guinea, and the Department of Territories for photographs and information in chapters 4 and 7; Iliffe Industrial Publications, London, for permission to quote from the *Science Journal* in chapter 1; the Bureau of Agricultural Economics, Canberra, for permission to quote in chapter 15; the Australian Mutual Provident Society for permission to quote in chapter 16 and for photographs in this chapter; the Japanese Consulate-General, Sydney, for photographs in chapter 5; the United States Information Service, Sydney and London, for photographs in chapters 8 and 17; the Danish Embassy, London, for photographs in chapter 10; the New Zealand High Commission, Sydney and London, for photographs in chapters 11 and 14; the Australian News and Information Bureau, Sydney, for photographs in chapters 4 and 13; Mr Eric Stockdale of Newington College, Stanmore, for assistance in the early stages of planning, and Mrs Joan Ladley and Mrs Phyllis Wheeler for their help throughout.

List of Maps

Contents

Part One

Introduction

1 The Geography of Agriculture

There is a growing fear of an impending world food crisis. More than half the people in the world suffer from hunger or malnutrition, and the population is increasing faster than the food supply. Some writers give statistics of food and population growth which show that a food crisis is imminent. . . .
It is estimated that in the next 30 years world population will reach 7,000 million—double the present size. To abolish hunger in the present population the food supply would need to be nearly doubled. To provide sufficient for 30 years hence, the supply would need to be increased by between 300 and 400 per cent, the biggest increase being in protein-rich foods.

1. The World Food Problem

Lord Boyd Orr, who made the above statement in the *Science Journal* in May 1968, was the foundation Director General of the United Nations Food and Agriculture Organisation from 1945 to 1948. He declared in his autobiography *As I Recall* (1966), 'The hungry people of the world wanted bread and they were to be given statistics.' Despite the implied criticism, the FAO has become the body concerned with the world's most pressing problem—how to feed its people. Statistics emphasise the problem dramatically.

In 1969 the world population totalled some 3,500 million. Of this figure about 2,300 million must be regarded as undernourished. This hungry majority has diets ranging from those with a relatively mild protein deficiency to those of no better than starvation level. The basis of these figures is the calorie per day and protein per day intake. North American and Western European populations consume averages of 3,000 calories and 100 grams or more of protein daily. There is even concern that these intakes are more than that necessary for optimum health—certainly many people consume far more than their energy requirements.

On the other hand, the populations of all or parts of Ceylon, India, Pakistan, the Philippines and non-European Africa have dietary averages around 2,000 calories and 30 to 50 grams of protein daily. Thus, in this sector of the world alone, over 700 million people have an average of only two-thirds of the necessary calorie intake and less than one half of the protein intake deemed necessary by FAO experts. The research by the FAO indicates that throughout the world over 500 million people have too little dietary energy, and a further 1,500 million people too little protein for normal health and activity.

In the present situation, 35 per cent of the world's population is adequately or more than adequately fed, while 65 per cent is suffering various degrees of starvation. It is estimated that to abolish hunger in the present population the food supply would need to be nearly doubled.

Staggering in magnitude though the problems of the present appear, those of the future at times seem unthinkable. Of the estimated population of 7,000 million in the year 2,000, the undernourished peoples of the world will number more than 5,000 million, or 75 per cent of the world total. With two out of three people starving in 1969, the prospect in 1999 becomes three out of every four, even if the present rate of food supply is doubled—that is, if every acre can be made to yield twice as much or if previously waste or empty land can be brought into production. Because of the alteration of the proportion of nourished to undernourished in the population, food supplies will need to increase by between 300 and 400 per cent on present production figures to provide an adequate diet for all the world's peoples.

In the light of such an overwhelming challenge, any study which examines the production of foodstuffs attains special importance—and the human science which embraces such study is agricultural geography.

2. The Scope of Agricultural Geography

Strictly defined, the term 'agriculture' is 'the practice of cultivating the soil in order to produce crops', and is clearly differentiated from 'pastoral farming', defined as 'the practice of breeding and rearing certain herbivorous animals', in a modern Dictionary of Geography. In 1936 the American agricultural geographer Derwent Whittlesey used 'agriculture' to describe both cropping and grazing, 'for lack' (as he put it) 'of a more fitting word'. It is in this broad application that the term is used here to include the rearing of animals and the growing of plants, as well as to recognise that activities like poultry-farming, oyster-farming and brackish water fish-farming are either agricultural industries in their own right or at least make use of many characteristic agricultural practices. With this in mind, there are certain comprehensive fields of endeavour which can be included in any definition of agriculture.

Included here are all plant products yielded for direct or indirect consumption as the result of a programme of seeding, cultivating and harvesting, and all animals or animal products resulting from a programme of breeding, rearing and fattening. A broad consideration such as this allows the inclusion of all crops which are consumed directly by humans or fed to animals and indirectly consumed as meat or animal protein. Oysters, mussels and the like, which are systematically 'planted' in frames, culled during the growing period, protected from many predators and finally gathered, can be included in a study of agriculture; but the haphazard gathering of beche-de-mer, abalone, toheroa and other marine creatures must be excluded. Of course, difficulties can still arise: grouse, partridge, pheasant and other game birds are hunted but often they are also farm-reared stock. Despite the hunting aspect and the isolation of all forms of hunting from the scope of agriculture, man is here involved in the deliberate attempt to produce the livestock which are farmed before being liberated.

An inflexible approach will not work, for agriculture is dynamic. For example, some fish being hunted by wasteful line fishing, trap or net methods at present seem inevitably destined to be bred and raised for market in enclosures of the ocean. Suggestions like 'fencing' areas of the ocean by electric fields and using dolphins as shepherds may not be so fanciful in the near future, for agriculture as Whittlesey saw it and as it has now developed to the present may be vastly changed in the last half of this century. As the world strives to feed its spiralling population, agriculture is recognising methods of food production that were outside its scope years ago. Attention to the production of certain single cell pond algae as high protein additives to diets, oilseed meal producing Indian multi-purpose food (IMPF), the grazing of wild beasts such as the African eland on poorer rangeland than would support conventional beef cattle, are all examples of the dynamic character of modern agriculture.

Apart from food, agriculture supplies many other necessities as the result of cultivation of plants and the rearing of animals. Chief among these are plant and animal fibres. Wool and cotton are the major examples of natural fibres, while a group of synthetic threads based on cellulose is playing an increasing part in the production of fabrics. Nor are the synthetics feared by primary producers to the same extent as two or three decades ago; today the tendency is to increase the variety of combinations of man-made and natural thread, instead of direct competition between natural and synthetic fibres.

Building and construction material ranging from lumber to plastics produced from cellulose is another main branch of finished goods resulting from agriculture. Increased research has led to a variety of building and insulating materials processed from the waste of plants grown for other purposes—mainly for their edible parts. Bagasse, the waste pulp from crushed sugar cane, is processed into a number of multi-purpose wallboards; sawdust from both hardwood and softwood plantations makes plastics, newsprint and construction materials. The inclusion of forests as an aspect of agriculture is perfectly justifiable if the tree is regarded as a crop—and individuals as well as public authorities are inclined more to this view in recent years.

There is a huge number of other industrial raw materials which result from agriculture. Casein from milk for pliable plastics; fertilizer, glue, gland extract drugs, bone-char and leathers from farm animals; drugs, dyes, perfume bases, insecticides, flavouring and oils from certain plants—these are only a few of the complex of raw materials flowing in increasing amounts as the main product or by-product output of farms.

3. Improving Traditional Agriculture

What must be done in order to bridge the chasm that lies between food production and the minimum consumption requirements? Because for the first time man has the technical ability to produce sufficient food for mankind, the view of traditional agricultural methods must be drastically altered. Although there are tracts of the earth's surface which can be brought under the plough, the overcrowded world obviously lacks additional arable acreages. The common view that India's main concern is a shortage of arable land is also held to be inaccurate, however, for it is not the lack of arable land that is the chief problem, it is the low degree of efficiency of agriculture. In all or parts of India, Cuba, Taiwan, West Malaysia, Egypt, Venezuela, Peru and many other countries, the level of food production per capita is little above the 1938 levels, in some places it may even be below. These are the areas of traditional farming, which in agricultural terms means production for subsistence.

If social problems can be left aside for the moment, the areas of subsistence agriculture can be revolutionised by the introduction of modern methods. The introduction of fertiliser programmes, or incorporation of rotation schemes and legumes where applicable to the farm, can be major steps. Mechanisation and the extension of rural power facilities can lift production levels on the labour input basis at least. Protection of crop and stock by weedicides, pesticides, shelter and fencing will lift levels on an acreage basis either through increased yield or a saving from damages. Improved marketing and transport facilities, while not making startling increases possible generally, will at least level out some of the glut and famine fluctuations so often experienced in the traditional farming lands. Possibly the most dramatic achievements, outside the introduction of irrigation water, result from the substitution for traditional grains and livestock of scientific and genetically high yielding species which may perform several times better than the traditional varieties.

Unfortunately, traditional methods of production often go hand in hand with traditional social structures which inhibit or completely prevent the introduction of progressive measures. India is the most obvious large scale example. In 1962 the animal population of India was estimated at 336 million, of which over 200 million were sacred cattle, over half of them male animals. These cattle, sacred to the huge Hindu majority, are to blame directly for some of the ills of Indian agriculture, and are symbolic of several other problems. Many beasts are of uncertain ownership, the bulk of them roam the countryside, suburbs and cities in their uncontrolled millions. They trample and ruin crops, break through flimsy fences and prevent the growth of seedling trees and the regeneration of grasslands. They are permitted to consume food desperately needed for human consumption, and in return they may not be eaten themselves. The milk output average of these haphazardly bred cows is the lowest of any in the world, averaging less in a year than reasonable quality milkers produce in a month in the dairy nations of western Europe. Calves are sacred too, and all bull calves reach maturity and add to the interbreeding and uncontrolled pattern which prevents any improvement in progeny.

These are direct hazards. Probably of greater significance even than this is the attitude towards cows as gratuitous fuel suppliers. In this country where less than 40 per cent of power is produced from fossil deposits of coal, gas and oil, cattle-dung makes up 30 per cent of the fuel supply, and a substantial reduction in cattle numbers would cause complete chaos in the fuel economy. Yet Indian soils lack the fertility levels necessary to maintain, let alone increase crop yields. It is estimated that 600 million tons of dung are lost in the grazing areas, and 400 million tons are burnt for fuel—a staggering 1,000 million tons representing the greatest wastage of agricultural substance in the world.

In similar fashion, a combination of traditional social and economic pressures combine to inhibit agricultural change in many countries of the underdeveloped world. The methods of changing the patterns of traditional agriculture must attempt to bypass social problems and concentrate on overcoming subsistence methods and replacing them with modern commercial practices. Nor are individual innovations sufficient to ensure success. There is an obvious advantage in introducing fertilisers, but better seeds and protection from disease and pests must also be part of the deal if full benefits are to accrue.

The problems of world agriculture which the United Nations Food and Agriculture Organisation is attempting to solve with the assistance of many other agencies, require all conditions to be favourable if successful innovations are to be introduced. If there is a resistance to change great enough in any of the fields of influence—social, political, cultural and economic checks, for example—then the desirable change in the agricultural structure will not be implemented even in its formative stage.

Bibliography

Alexander, J. W., *Economic Geography*, Prentice-Hall, Englewood Cliffs, N.J., 1963.
Chisolm, M., *Rural Settlement and Land Use*, Hutchinson, London 1962.
Laut, P., *Agricultural Geography*, Thomas Nelson, Melbourne 1966.
Moore, W. G., A *Dictionary of Geography*, 4th edition, Penguin Books, Harmondsworth, Middlesex, England, 1968.
Orr, Lord Boyd, *Science Journal*, Vol. 4 No. 5, May 1968, Iliffe Industrial Publications, London.
van Royen, W., *The Agricultural Resources of the World*, vol. 1 of *Atlas of the World's Resources*, Prentice-Hall, New York 1954.
Whittlesey, D., *Major Agricultural Regions of the Earth*, Annals of the Association of American Geographers, 1936.

2 Classification in Agriculture

Despite the great variety of products resulting from the cultivation of plants and the rearing of livestock, the basic aim of agriculture is the production of foodstuffs. The methods under which crops are grown and animals reared vary widely in different economies, and the task of the agricultural geographer is to classify the various forms. Early attempts at classification of agriculture in this century have been based on either a regional or topical study.

1. Topical Studies

The topical concept deals with a single commodity and considers all aspects of its production, processing and marketing, and it is often set in the several world areas which contribute towards the total supply of the product. There may be combinations of the two, where a topical study is carried out within a specific region. Topical studies are often important in western European countries where the processing of raw materials from areas outside Europe takes place. British geographers have produced comprehensive studies of agricultural products originating in the tropical areas, for example; the details include not only methods of production in the country of origin, but the analysis of transport, processing and retailing of the finished goods.

2. Regional Studies

The regional or area concept involves complex considerations and much theorising by geographers, and the almost interminable discussion as to the nature of the region has unfortunately also resulted in some rather confusing definitions. The geographic region is delimited from its surroundings, and contains within its boundaries an area of distinct identity or 'character'. The problems of delimitation can be difficult, but there is a number of commonly used bases—for instance, landform structure or climatic differences, differences in, or differences of stress on, certain land uses; economic and social differences; or combinations of these and other variables. The region is thus delimited as the result of a selection of criteria by the geographer. If he chooses certain measurable characteristics which are present in some part of the earth's surface or either absent completely or absent to a degree outside this part, then boundaries can be drawn. The extent of the region is governed by the statistical limits the geographer chooses to impose. The regional study is common for example in Australia and the United States of America, countries where explanations of the character of certain areas are desirable for regional planning purposes. A great deal of statistical investigation is carried out when regional plans are drawn up.

There are differences between the scale of studies, too, ranging from the intensive study of quite small areas up to a general study of quite large areas. The detailed study of the small area has developed the greatest refinements in technique, but quite a few of the larger area studies are too generalised to be really meaningful.

3. Early Attempts at Classification

Some of the earliest work on a relatively small scale but covering quite large areas was that of the American geographer O. E. Baker. With V. Finch he produced the first major world study, *The Geography of the World's Agriculture*. This was an atlas, with maps based on statistical information, accompanied by descriptions of areas of production, output and marketing of the main agricultural commodities. From 1925 to 1940 a number of continental studies and large area studies followed. Baker himself contributed the classification of United States agricultural regions, and others—Griffith Taylor (Australia), C. F. Jones (South America), van Valkenburg (Asia), H. Shantz (Africa) and Jonassen (Europe)—submitted studies which formed a series in the journal *Economic Geography*. As well as these small scale, highly generalised studies, there were several more detailed regional studies, for example by Durand on dairying regions and by Baker on the Great Plains, and topical studies by J. C. Weaver on barley in the United States and by G. B. Shaw on swine production.

During the 1930's, while the above series was being completed, new definitions of agricultural categories were being attempted. W. D. Jones proposed a cultivation index, the measurement of occupance in agricultural land by comparisons of the ratio of land under various crops; a specialisation index, obtained from the percentage of crop land under individual crops; livestock units per 100 acres of crop land as a measure of relative importance of animals or livestock; ratios of output, such as gallons of milk to the crop acreage; and detailed relationships between work animals to total livestock. Such quantitative work permitted the construction of isoline maps, and from these Jones defined three categories of agricultural activity—commercial grain farming, commercial livestock farming and commercial dairy farming. Jones was later (1935) to be co-author with Derwent Whittlesey of a map of world agricultural types, based on Whittlesey's classification of agriculture.

Later, Baker and R. Hartshorne were to employ quantitative measurements for further development of the specialisation index. In the United States dairy areas, for example, Hartshorne measured the gallonage of milk produced per 100 acres of crop land, thus establishing a critical figure for delimiting a boundary based on this criterion. Baker used such criteria as percentage of land in crops which allowed him to delimit the cropping as opposed to the Great Plains ranching region; bushels of maize per square mile to delimit the boundary of the Corn Belt; ratio of values of vegetables/fruits per square mile to the value of all crops per square mile for the boundary of the Atlantic coast Truck Farming region; and ratio of arable land acreage to hay and pasture acreages to delimit the Hay and Dairy belts. From these single criterion studies other workers have developed multiple criteria approaches, such as combinations of livestock and crop units.

Other quantitative measures have been used with varying degrees of success. Per capita gradings have been used by Hartshorne by relating quantities and types of production to people; value of production has been used to allow comparisons of different types of activity. These are not used as much as the crop land to farm area ratios, the statistics for which are relatively easy to obtain.

4. Whittlesey's 1936 Classification

Although he invited criticism of his work, Derwent Whittlesey produced a classification entitled *Major Agricultural Regions of the Earth*, which is to agricultural geography what Koppen's classification is to climatic geography. It is based on five criteria:

(a) the association of crop and livestock;
(b) the intensity of land use;
(c) the methods of production;
(d) whether commercial or subsistence, depending on the disposal of the products;
(e) the farm capital revealed in structures and buildings.

On the basis of these five criteria, Whittlesey's classification has thirteen categories, summarised below.

i. Nomadic Herding
This type occupies areas too dry for cropping and is the only form of subsistence grazing. Stock commonly include sheep, cattle, goats, camel or reindeer. The length and route of the migration is a response to amount of forage and in particular to availability of water. Animals provide nearly all the needs of the owners. Little statistical evidence is available.

ii. Livestock Ranching
A commercial sedentary form found mainly in the semi-arid and arid lands. Modern methods, especially efficient transport, are necessary; markets, often located overseas, are vital. There is usually a low ratio of animals—sheep, cattle and goats—per acre and little cropping. This type overlaps the areas of Nomadic Herding, employing the same landscape.

iii. Shifting Cultivation
A low latitude subsistence type, based on soils ranging from rain forest laterites to park savanna types and oceanic deltas to mountain basins over 14,000 feet in altitude. Few animals are kept except poultry, pigs, goats and cattle (outside the tsetse fly areas only). Poor soils leach rapidly, creating a need to move every few years, until after a time the fields become too remote from the village and it too must move. Small tribes using crude tillage methods are usual, no statistics are available and few studies have been made.

iv. Rudimental Sedentary Tillage
Among the shifting cultivation areas are some sedentary groups where populations may become rather dense, especially on small islands, mountain valleys and in areas of favourable soil and climate. Because they cannot or will not leave these areas they re-use the plots, although fertiliser use is low, tools and knowledge poor. Some stock may be kept and long periods of bush fallow are common. Influence from outsiders may have encouraged sedentary habits; certainly European demand for tropical crops—cacao, oil palm, coconut, rubber, peanuts and cotton—has assisted the primitive farmers to grow these and other market crops among their subsistence types, and in a few areas the agriculture has developed to resemble the highly developed oriental type. Difficulties in mapping except along major transport

routes and around towns and cities are fairly obvious in these areas. Some household equipment and more substantial housing distinguish these groups from the shifting tribes, but statistics are few and studies limited.

v. Intensive Subsistence Tillage with Rice Dominant

In humid south and east Asia, two types of intensive subsistence tillage, with or without rice in the pattern, are found. Rice is included where the season is long enough and creates the densest population areas on the earth. Padi areas—irrigated deltas, floodplains, coastal plains and terraces—produce more grain per acre than any other crop, with two harvests in hot climates (resulting in more than 100 per cent of the land being used) and one where a cool season is found. Within these areas away from irrigation, various other crops—grains, oilseeds, cotton, tea, pepper, mulberry trees—are found. Freshwater fish farming and waterfowl are a vital part of the pattern.

Hand labour, except for ploughing, is common and tools are crude. Flooding, natural or irrigation, helps to grow two crops. Interculture, rotation, and fertilisers provide high yields per acre, but this still only allows low returns per capita of population.

vi. Intensive Subsistence Tillage without Padi Rice

In areas bordering the previous type (number *v*), where a lack of moisture or a short growing season occurs, farmers are forced to substitute several cereals (none of them as productive) for rice. These regions are more generally located towards the interior. Vegetable and tree crops are much the same, with irrigated plots, unirrigated fields and slopes, and much the same congestion as their neighbours, although population densities are less. Greater use is made of technology and transport to alleviate famine in some areas. Statistical methods can distinguish between this type and the previous classification.

vii. Commercial Plantation Crop Tillage

This is a European superimposition in a few areas of the types numbered *iv* and *v*. This type is heavily dependent on outside markets mostly in mid-latitudes and both capital and staff are drawn from Europe or North America. This in turn creates a demand for imports of machinery, fertiliser, flour, dried foods and canned goods. Usually field hands are local, although some may be brought from other areas. The plantation is a device to ensure commercial production of tropical crops—for example, rubber, sugar and tea. Growing industrial populations outside the tropical areas have created markets and the sophisticated technology of these consumer areas has made possible many refinements in plantation methods. Refrigeration has extended the range of crops able to be produced for export, such as bananas; improved machinery has revolutionised processing for sugar and rubber; other products have met industrial demand—rubber, sisal, copra—while new uses were found for cacao, cotton and vegetable oils. The traditional plantation is now losing its character, as more and more of the large holdings are broken up into small units and rented to share farmers. Moreover, all plantation crops can be grown successfully by native small holders if no processing is required before marketing; thus there is always the danger of gluts of the particular crops of the region and constant overproduction, unless consumer demand takes a sudden upsurge.

viii. Mediterranean Agriculture

This type represents one of the most dependable traditions of land use, with methods so tried and familiar that even in the strange continents of the New World settlers struggled to develop regions which seemed to echo the climate and landforms of established agricultural areas of the Old World. The unique Mediterranean environment with its association with mountains and hills has created a distinctive livestock-crop relationship of:

(i) winter crops grown with rain;
(ii) all year crops grown with rain;
(iii) all year crops and summer crops grown with irrigation.

Subsistence and cash cropping exist side by side on many farms, with the choice of product dictated by the rainfall of the region; for example, North Africa concentrates on goats and barley, while South Europe prefers sheep and wheat. Other influences on choice of product include tradition, market opportunities and government planning or protection. Methods are intensive, varying only to accommodate differences in land tenure in different regions or to adjust to specific slopes or climatic conditions. Settlement patterns are traditional. Land values are high, and thus homesteads are built on poor hills where they will not interfere with the cultivation of good garden soil. With small plots producing sufficient income to support a family, nucleated settlements are the rule.

ix. Commercial Grain Farming

In interior regions, where the main crop, wheat, comprises as much as half the total farmland, progressive methods employing much machinery are usual, although return per acre is low compared with more intensive tillage areas. These marginal areas suffer from erratic and unreliable rainfall and the farmer has no predictable income from season to season. Although nothing can be done to alter the vagaries of the climate in these areas, consistent research into new strains of wheat has developed hardier types and allowed the limits of these regions to move further towards the poles.

x. Commercial Livestock and Crop Farming

Often termed 'mixed farming', this type features crops and livestock in varying ratios. The principal crops include wheat, maize, oats, rye and barley, with root crops such as potatoes, turnips and sugar beet. Animals include hogs, cattle (both beef and dairy) and poultry, and the main cash products come from livestock. Methods are progressive with the use of much machinery, well established rotation and fertilisation patterns, and the recognition of the importance of stock breeding and selection. This is a highly efficient form of agriculture, bringing the farmer high returns.

xi. Subsistence Crop and Livestock Farming

This type occupies the poorer marginal lands on the borders of commercial grain areas, or smaller pockets of poorer soils which are backwaters of development. Returns are not high and in contrast with the previous classification (number *x*) methods are rarely efficient and use of machinery and fertilisers is less progressive, even though much the same crops and livestock are used.

xii. Commercial Dairy Farming

One of the most efficient of all agricultural types, this form is found where urban markets are created. The returns on these farms are very high, but capital establishment costs are great for housing, elaborate machinery and fodder storage. The differences between dairy areas and neighbouring crop and livestock regions are established by ratios of hay to grain, the high percentage of dairy cows to livestock of other types and the high population density of the dairy belts.

xiii. Specialised Horticulture

When horticulture occurs on a large scale, it is automatically specialised into areas where conditions are particularly favourable. Most of these areas are too small to be shown effectively on a world map. Each region concentrates on the few crops which are best suited to prevailing conditions, and the biggest demand comes from a local market created by a large manufacturing or urban population. With market gardens so close to big cities, the aim must be maximum cropping of minimum acreage and the result is that market garden land has a higher value per acre than any other agricultural land.

Bibliography

Alexander, J. W., *Economic Geography*, Prentice-Hall, Englewood Cliffs, N.J., 1963.

Chisholm, M., *Rural Settlement and Land Use*, Hutchinson, London 1962.

Fryer, D. W., *World Economic Development*, McGraw-Hill, New York 1966.

Thoman, R. S., *The Geography of Economic Activity*, McGraw-Hill, New York 1962.

van Royen, W., *The Agricultural Resources of the World*, Vol. 1, Prentice-Hall, New York 1954.

Whittlesey, D., *Major Agricultural Regions of the Earth*, Annals of the Association of American Geographers, 1936.

Part Two

Subsistence Agriculture

3 Subsistence Agriculture

1. Introduction

Primitive subsistence agriculture is practised in rain forest regions and surrounding areas. The cutting down of vegetation exposes the lateritic soil to the sun and the rain, and leaching and erosion results. Because of the poor soils in these areas, the primitive indigenes have to move their garden plots every second or third year, or even every twelve months in very poor areas.

When this happens virgin land is sought in forested areas nearby. However, as the plots become more remote from the village itself, the tribal group finds it expedient to move the village to a new site. A new section of the forest is slowly cleared, the village is set up, and the procedure of establishing small cultivated areas is commenced again.

Local conditions determine the frequency of movement of the village from one place to another. Comparatively good conditions could mean that the village is moved every ten to twelve years; on the other hand, poor local conditions probably necessitate movement more frequently.

In the better areas, where sufficient rain falls and the temperatures are high enough, padi rice becomes the basis of an agriculture which is so intensive as to enable the support of some of the densest populations on the face of the earth. Two crops of rice per year are common in areas where the climate is suitable. Deltas are irrigated, terraces constructed, and flood plains and coastal plains are utilised for the intensive cultivation of padi rice. When water is insufficient and the winter becomes too cold, then one crop of rice only is cultivated. Rice gives high yields and keeps well and as a result is a grain in great favour in these lands.

Man-made ponds also make their contribution: fish are taken from the rivers and allowed to mature to the stage when they can provide the protein which is so important in making a reasonably balanced diet. Water also supports myriads of ducks which feed on aquatic plants. These water plants can also be used in some areas as 'green' fertiliser.

Probably as much as twenty per cent of mankind is still living largely on a subsistence basis. Basically, subsistence agriculture is aimed at supplying the needs of a small group—the family or tribe in its simplest form—and the intention is to grow as much food as possible for local consumption. More sophisticated subsistence agriculture does provide for some cash sales, but the basis is still self-consumption. Commercial agriculture aims to supply an outside market, often international in extent, and is directed at large urban populations which produce no food for themselves.

Even though there are wide contrasts between the various subsistence types, several generalisations can be made.

THE WORLD
SHOWING AREAS OF
SUBSISTENCE AGRICULTURE

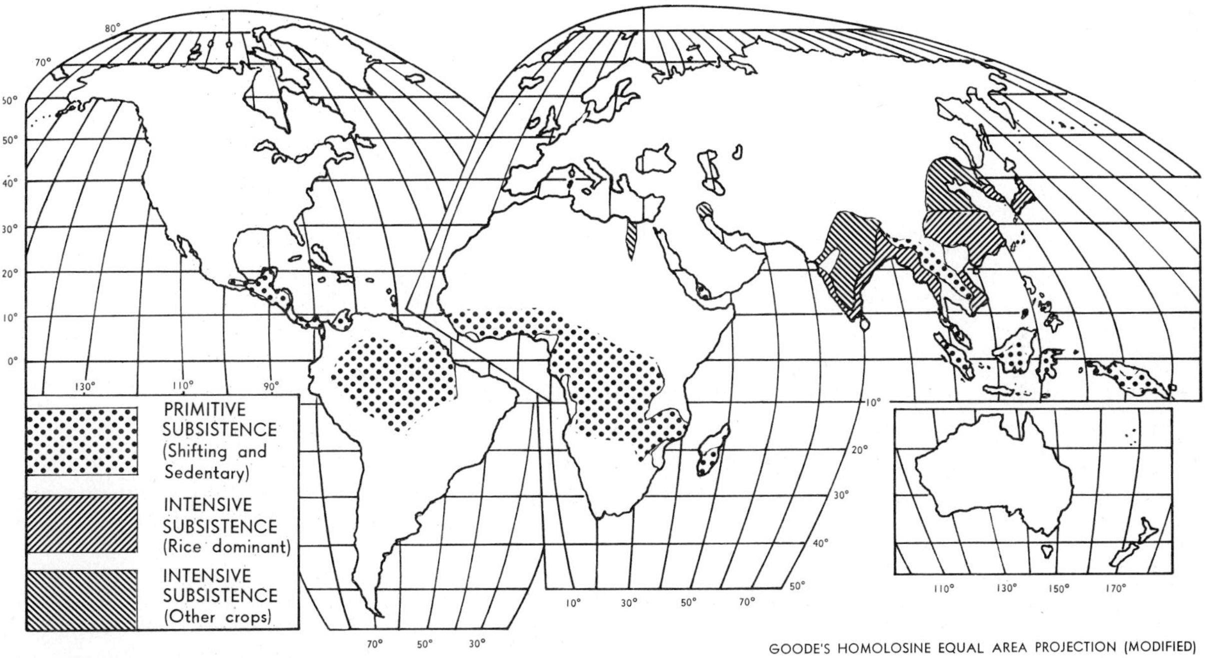

2. Basic Considerations

i. Location

Because of the scattered nature of the subsistence peoples, especially primitive types, it is difficult to map the extent of this agricultural type. Little research has been done in many instances. There is considerable overlapping with the last vestiges of hunting and gathering groups on the one hand and with commercial agriculturalists on the other.

The primitive subsistence groups—both shifting and sedentary—occur throughout the intertropical world. In South and Central America and the Caribbean islands they are seen from South Mexico to South Bolivia. Although they are found on both sides of the Andes, the greatest numbers occur throughout the Amazon Basin. Certainly the most primitive of the groups are seen in the isolated areas of Amazonia, along the great river Amazon and many of its tributaries.

In tropical Africa there is a similar scattering. The northern limits are in the Sudan savanna lands and the extent is south almost to the Republic of South Africa. The least advanced are in the basin of the Congo where isolation has preserved much of the character of this ancient form of agriculture.

The Far East and South East Asia has, numerically, the greatest concentration of subsistence agriculturalists. Non-rice Northern China, Japan and India-Pakistan account for a very large share of intensive subsistence people. Most South-East Asian countries and the numerous islands south and east of Asia have examples of both primitive and intensive agriculturalists.

Just as it is difficult to separate the various forms of subsistence agriculture, so it is not easy to generalise on specific locations within the areas outlined. The range is from coastal deltaic plains in mainland east Asia to mountain basins in India over

10,000 feet in elevation; from the humid wet jungles of central Amazonia to the savanna periphery of the Sudan.

ii. Climate

The correlation between subsistence agriculture and tropical climate is a strong one, although the great latitudinal spread and altitudinal range makes temperature generalisations invalid in many instances. Rainfall everywhere is moderate to high. In the tropical lowlands the growing season is all year—frosts are unknown and around 70°F. is a minimum average monthly figure. The northern edges of the subsistence areas, however, may have up to five months of frost, as in mainland China. Similarly, in the highlands temperatures can restrict plant growth in one third of the year. The driest areas of intensive subsistence agriculture—parts of mainland China, for example, receive around 20 inches per annum, although most areas probably experience between 40 and 80 inches per annum. The primitive areas are generally located at the humid end of the scale and mostly experience more than 40 and up to and over 100 inches per annum.

Seasonal variation is great in such an extensive zone although the summer monsoon and its influence is common. Reliability of precipitation is high in most instances, except along the savanna boundaries and where there is an occasional failure of the monsoon, or when rainfall is excessive. This is highly significant when immense populations are involved, and examples of flood and drought in India and China have been catastrophic.

Selected Statistics (temperatures in degrees F.; precipitation in inches)

Manaos, Brazil (3° 10′S; 60° 0′W.)

	J	F	M	A	M	J	J	A	S	O	N	D	Year
Temp.	80	80	80	80	80	80	81	82	83	83	82	81	81
Precip.	9·2	9·0	9·6	8·5	7·0	3·6	2·2	1·4	2·0	4·1	5·5	7·7	69·8

Zanzibar, East Africa (6° 0′S; 39° 20′E.)

	J	F	M	A	M	J	J	A	S	O	N	D	Year
Temp.	83	83	83	81	79	78	77	77	78	79	81	83	80
Precip.	2·8	2·2	6·1	14·1	10·7	2·0	2·4	1·7	2·1	3·6	7·5	5·0	60·2

Bandoeng, Indonesia (6° 55′S; 107° 40′E.)

	J	F	M	A	M	J	J	A	S	O	N	D	Year
Temp.	72	72	72	72	72	72	72	72	72	73	72	72	72
Precip.	7·6	7·1	9·6	9·0	5·2	3·6	2·6	2·3	3·6	6·7	8·9	8·5	74·7

iii. Topography and Soils

Where intensive rice farming is dominant, the preferred topography is the coastal delta lands and valley floors where relatively flat land can be irrigated. This does not preclude sloping land, however, as the many examples of terraced hillsides throughout South-East Asia testify. The more primitive peoples accept hillside sites where possible, for here drainage and disease problems are less. On the other hand, the erosion of these slopes creates difficulties too, so that plots are often forced into valley bottoms.

Soils everywhere in the forested lands suffer organic and nutrient deficiencies. As the slash and burn technique is the only practicable method of clearing virgin lands, the organic material which has accumulated so slowly is partly destroyed. Certainly potash from the burnt litter adds to the topsoil fertility, but its light nature makes rapid erosion also possible.

Nutrient deficiencies are the result of leaching. In this process, the soluble salts are carried to the lower horizons by the steady percolation of rain water. Crop plants are generally too shallow rooted to take advantage of the zone of accumulated nutrients. At the same time a process of eluviation takes place as the fine clay particles are physically transported to a zone of concentration, the 'gley' layer.

Thus the typical laterite develops. Coarse, sandy and relatively infertile, often stained bright red by iron oxides, these soils present a severe handicap. Clearing and burning speeds up the laterisation process and creates the necessity of shifting in a few years to a new site.

By contrast, many areas of intensive subsistence agriculture are located on relatively fertile, and in some instances highly fertile silt soils. Continuously cropped for centuries, the soils of China's great river basins demand continual fertilising. This occurs naturally to a degree with the silt-enriched flood waters that give the Hwang-Ho the title of 'Yellow River'. In addition, any organic material—spent crops, pond slime, aquatic plants and limited quantities of green manures—are employed. Animal dung is limited in quantity, which means that human excreta is universally used in most of the rice world. Pigs, ducks, geese and the padi-grown fish add not inconsiderable quantities of manure where they are kept intensively. In Japan even silk-worm excreta is spread on the fields.

No study of rice growing regions would be complete without reference to the terraced hills typical of much of South-East Asia. For over forty centuries in the Philippines terraces have been built—thousands of miles if laid end to end. Often the retaining stone and soil dikes are one third as high as the padi is wide. Streams are tapped at high points to lead into the padi, and from tier to tier of dikes the water is led to the lowest level.

3. Types of Subsistence Agriculture

Subsistence Agriculture can be conveniently grouped into four categories:

(i) Shifting Subsistence Cultivation;
(ii) Sedentary Subsistence Cultivation;
(iii) Intensive Subsistence Cultivation (Rice-dominant);
(iv) Intensive Subsistence Cultivation (Non-rice).

While some characteristics already outlined indicate similarities as regards the intention to produce for the indigenes' own consumption, each type has considerable variation from the other.

i. Shifting Subsistence Cultivation

This is the most primitive form of agriculture, and is but little removed in degree of sophistication from hunting and gathering. Indeed, in many instances the agriculture is practised in conjunction with greater or lesser amounts of hunting, fishing and gathering.

a. Location. These primitive peoples are restricted to inhospitable locations in isolated forest areas of South America, Africa and parts of South and South-East Asia. Few groups today have not had some contact with Europeans, or at least from other tribes who have made contact with civilisation. There are, however, notable cases which still reveal a way of life little removed from Neanderthal forebears. Isolated hill tribes in New Guinea have only been contacted in the last couple of decades. Interior Amazonia possibly still has groups who have not seen foreigners and this was true in parts of Central Africa in the early part of this century.

b. The natural environment. Since these areas are located in the warmer sections of the low latitudes, frosts are generally unknown. Conditions are wet all year, often with a summer concentration of rainfall. It is possible to grow crops for the greater part of the year, but cultivation techniques usually mean a limited number of cereals or tubers; manioc is most usual, and a few introduced plants—beans, peanuts and salad types, for instance.

The natural vegetation ranges from true rainforest, through lighter forest types, bush and thorn to savannas—reflecting the gradation from wetter to drier areas. These vegetal types are located on lateritic red soils or lighter red and yellow laterites and podsols. These last types are generally more fertile, since leaching is not so advanced, but they are low in organic content. The burning of the cleared soils increases leaching and the soils deplete rapidly.

Drainage depends on slope but increased slope speeds up the erosive process, especially early in the growing season. This compounds the problems of yield and length of continuous cropping. One season of cropping only may result and the maximum is probably no more than three or four.

c. The cultural background. These groups practise the slash-and-burn technique and rely almost entirely on crops. Included here are staples like manioc, sweet potato, maize or millet, yams, sugar cane and other annual vegetables. Yields are low both per man and acre, the lowest of all the agricultural types.

The forest supplies all needs, except for the occasional traded luxury. The social unit is small—family or tribe with a simple division of labour: men select and clear the site and may plant the crop, women cultivate and harvest the crop. The dependence on the annual crop is great, and starvation or near starvation is characteristic between seasons or in drought years.

Implements are crude. Digging sticks and wooden hoes are still used in many tribes. Ploughs are practically unknown, and animals are not used in any case, so tools are made for man only.

Similarly tribal life is primitive. Possessions are few and household items limited to a few trade goods. When some gathering of forest products for sale or barter is carried out, the people are a little better off. The move to a new village site is at least relatively simple since the forest provides all materials for dwellings.

Few contacts are made with other groups or foreign peoples. This is of course largely the cultural reason for lack of progressive techniques. So much time is consumed in food producing that advance must be limited, yet the numbers of primitive shifters continue to decrease. Migration from other countries and other areas continues to make inroaus into the shifting subsistence areas. The limiting factor is subsistence itself. Lack of cash opportunities inhibits advance. Isolation prevents education. Illiteracy is total and even simple agricultural techniques remain unlearned.

d. Shifting subsistence cultivation practice. Availability of satisfactory sites is a problem even though there are large tracts in the rain forests. Tools are primitive to the degree that severely handicaps clearing of large trees. The coarse savanna grasses are not easy to clear and burn except at the height of the dry period. Thus the density of peoples is very low, except in the better environments. As the yield per man and per acre is low so the density of population is restricted. With consequent lack of social intercourse the chances of advancing techniques decrease.

ii. Sedentary Subsistence Agriculture

Generally, similar environmental conditions exist in this type. There are a number of differences, however, which indicate the relative advances that these peoples have made.

Climatic conditions and thus vegetation and soils may be superior. Where conditions are drier soils will generally have higher fertility. This may permit a fully sedentary practice or a modified semi-shifting type. In the former cases better drained lowland soils in river valleys may have a comparatively high degree of fertility. If this is maintained by rudimentary fertilising techniques, and the areas are relatively large, the tribes have every possibility of establishing permanent villages.

Among the semi-shifting peoples the village may remain for perhaps one or two decades. The cultivators work at increasing distances from the village until it becomes too time-consuming to travel to the plots. It is only after quite a few years that a new village site must be selected.

Because of the above factors and an increase in social contact the peoples may make reasonable advances in agricultural technique. There may be an advance to the point which allows some crop surpluses for sale. With this limited cash, trading permits the purchase of improved implements. In other cases gathering of forest gums, waxes, dyes, rubber and other natural products or some growing of cash products occurs.

Political stability, the growth of colonialism and modern nationalism may make the step from shifting to sedentary cultivation easier and more rapid. Government and private organisations, foreign teams of agricultural advisers, missionaries and so on may help to give basic education and the chance of earning some currency.

iii. Intensive Subsistence Cultivation (Rice-dominant)

Vastly different in nearly all aspects is intensive subsistence cultivation, particularly when rice is the basic crop.

a. Location. It is easy to locate with a fair degree of accuracy the intensive subsistence rice areas. The principal region is the Far East, embracing most of central and south China, east Pakistan and the whole of the coastal fringe of India. South and East Asia are included: the coastal nations of the Malay Peninsula, almost all of Java, west Sumatra and the northern Philippines. In the north, Korea and Japan's south islands complete the general pattern.

b. The natural environment. The latitudinal range of subsistence rice growing does not preclude generalisation of the climatic factors. The two limiting considerations are the growing season and moisture. Rice requires four to six months free of frost, coupled with moderate to high monthly temperature averages—about 70°F. In conjunction an adequate supply of water is needed—80 inches per year or its irrigation equivalent for wet rice, and 50–80 inches for dry rice.

As the frost-free period lengthens, so too does the period devoted to rice. In the extreme, three crops of rice can be grown each year in the same field, and two crops are not uncommon.

Soils and vegetation in these areas reflect the influence of man. In the natural state the soils are quite fertile, being based primarily on floodplain silts and better alluvials. Where terracing has been carried out the soils may be poorer, but consistent attention to fertility has done much to change their natural character. The natural vegetation has largely disappeared from these lands, being limited to the higher and steeper areas.

c. Intensive subsistence cultivation practice. The methods of intensive subsistence are quite different from the primitive subsistence forms. Being wholly sedentary, and supporting huge densities of population, agricultural practices must aim at very high yields. This is achieved by several means, the chief being maintenance of soil fertility, and when possible double or treble cropping, interculture and the intensive use of all available space—ponds, channels and channel banks as well as land.

Several forms of manuring are used. Animals—draught oxen, pigs and poultry and water fowl—are very important in this respect. Nightsoil is in common use throughout the regions, and various forms of organic material—spent crops, seaweed, ashes and household waste—are composted or dug in to break down quickly. Nothing which can assist to produce the enormous yields of rice required for the populations can be wasted.

Not the least important are the fish which are raised in the padis. Carp have been grown for centuries in China, and there is great interest in new species like the African *tilapia*, a variety which grows to half a pound in four months and tolerates a wide range of water conditions. Fish very effectively convert refuse to valuable protein, and assist in maintaining fertility in the padi at the same time with their excreta.

These methods permit high per acre yields. Although variable throughout such a large region, from two to three tons per acre of padi rice is produced throughout the areas. Add to this the not inconsiderable production of other vegetables and food from fish, fowl and pig, and very high output is obvious. Unfortunately, the huge populations have a low average calorie intake. Often it is only 75 per cent of the 2,500 calories per day which dietary experts insist is necessary for mere maintenance of bodily health under conditions of manual labour in the tropics.

d. The cultural background. The population densities and pattern of settlement reflect startling population growth rates, and the pressures inherent in these population explosions. Farm sizes often average no more than two acres and average family size may be about five persons. It is well that rice is so productive, since population densities are anything from 1,000 to 2,500 persons per square mile. If the primitive subsistence agriculturalists are bonded to their way of life by isolation then it can be said that the subsistence rice growers suffer the same enslavement to their subsistence system because of too many people in a limited area.

Other cultural features help explain some of the difficulties these regions face. Millions are bound to restrictive diets by their religion. India's sacred cows soil and trample food which could feed thousands of peasants. The beasts occupy in total huge acreages of valuable land yet produce the lowest milk yields in the world. Traditional methods bind the peoples to their fields, while lack of cash prevents graduation to more efficient tools and improved techniques, seeds and fertilisers.

On the national scale only the spending of many millions of dollars in flood mitigation and conservation schemes will alleviate some of the pressures. Illiteracy, endemic disease and antiquated transport systems are all restrictions which prevent advance beyond the subsistence levels of living.

iv. Intensive Subsistence Cultivation (Non-rice)

It has been indicated that there is a fairly clear environmental boundary between the rice and non-rice areas. Where padi is not dominant the lower yielding dry rice may replace it. Other cereals—wheat, maize, grain sorghum and millet, together with a wide range of vegetables and fruit—are also grown.

There is less dependence on the amount of irrigation water but the intensity of land is no lower. Population density may be less, but when correlated with yields much the same pattern of basic low level subsistence persists.

Bibliography

Alexander, J. W., *Economic Geography*, Prentice-Hall, Englewood Cliffs, N.J., 1963.

Clarke, C., and Haswell, M., *The Economics of Subsistence Agriculture*, MacMillan, New York 1966.

Forde, C. Daryll, *Habitat, Economy and Society*, Methuen, London 1963.

Fryer, D. W., *World Economic Development*, McGraw-Hill, New York 1965.

Gourou, P., *The Tropical World*, Longmans, London 1959.

Highsmith, R. M. (Ed.), *Case Studies in World Geography*, Prentice-Hall, Englewood Cliffs, N.J., 1961.

Jones, C. F., and Darkenwald, G. G., *Economic Geography*, 3rd Edition, MacMillan, New York 1965.

Laut, P., *Agricultural Geography:* Vol. 1, Systems, Subsistence and Plantation Agriculture, Nelson, Melbourne 1968.

Robequain, G., *Malaya, Indonesia, Borneo and the Philippines*, Longmans, London 1958.

Thoman and Patton, *Focus on Geographic Activity: A Collection of Original Studies*, McGraw-Hill, New York, 1964.

Whittlesey, D., '*Major Agricultural Regions of the Earth.*' Annals of the Association of American Geographers, Vol. 26, 1936.

Questions

1. Clearly differentiate between the various forms of subsistence agriculture.
2. 'Subsistence agriculture is typical of agricultural systems in static societies.' Carefully analyse this statement.
3. 'Subsistence agriculture forms an integral part of the "cycle of poverty".' Critically discuss.
4. Although intensive subsistence agriculture (rice-dominant) involves a low level of farming technology, it cannot be said that the system is entirely unsuccessful. Why ?
5. Account for the decline in the number of people practising shifting cultivation.
6. Name the major areas where intensive subsistence (rice-dominant) agriculture is carried out in South-East Asia. In what ways does the most advanced form of subsistence farming differ from that of the least advanced subsistence agricultural practice ? Give reasons and examples.
7. One of the major characteristics of subsistence agriculture is that there is little or no surplus production. What are the economic and social ramifications of this ?

4 Primitive Subsistence Agriculture in New Guinea

1. Introduction

It is only in recent years that many New Guinea peoples have emerged from subsistence agriculture to the growing of some cash crops. In isolated, inaccessible areas primitive subsistence agriculture is still quite common. Although there were differences based on environmental conditions, the basic elements of native agriculture throughout Papua-New Guinea were similar. This section examines aspects of subsistence as they apply overall, with some specific references to groups in North-east New Guinea—primarily the Chimbu and Kup peoples, south of the Bismark Range. Here peoples living in the Wahgi and Chimbu river valleys and the valleys of tributary streams retain many aspects of ancient subsistence agricultural methods.

2. Location

The peaks of the Bismark Range, the surrounding peaks and the Southern Kuba Range rise over 12,000 feet with few major peaks below 10,000 feet. Most of the native population lives and works between the valley floor and the lower slopes of the lesser ranges—generally up to about 5,000 feet altitude. There is no hard and fast rule here though the basic consideration is a reasonably fertile soil without excessive slope. Even so, relatively flat land is only available in limited amounts. Centuries of cultivation have reduced areas of virgin forest and re-use of gardens is the normal pattern.

It is only in the most inaccessible areas that a 'pure' form of subsistence living is still practised. Most areas have had some contact with European missionaries or the Australian Administration, and in many areas traditional methods have been completely replaced by smallholder cultivation, plantation agriculture or a combination of both types.

3. Climate

Rainfall is generally well distributed and abundant especially in the south-west and in the Highlands themselves.

Associated with the high rainfall is a warm summer and cool winter. Variations are considerable when altitude and aspect are taken into consideration. The highlands can be quite cold, especially during May and November—the period of the 'Southeast' when less rain falls. As far as conditions for gardening are concerned, all year agriculture is possible in the warmer areas and the more sheltered highland valleys and hill slopes. Here temperatures during the daytime seldom become extreme, although nights are cold in the drier periods of the year. In the alluvial coastal plains conditions are similar to typical tropical forest lands—high humidity and low diurnal range are more obvious.

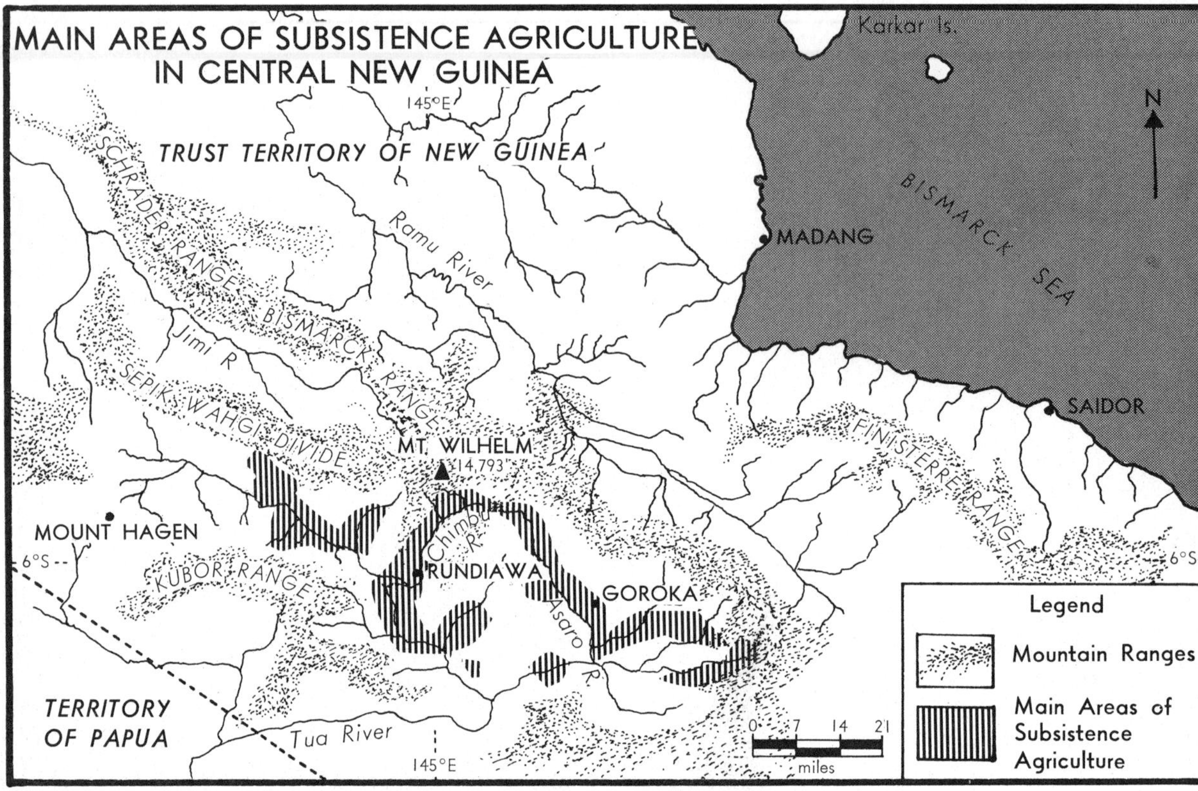

Selected Statistics (*temperatures in degrees F.; precipitation in inches*)
Goroka, Papua (6° 03′S; 145° 23′E.)

	J	F	M	A	M	J	J	A	S	O	N	D	Year
Temp.	69	69	69	68	67	66	66	67	68	68	68	69	68
Precip.	9·8	10·8	10·3	9·2	3·9	2·1	2·2	2·6	3·5	5·2	5·5	10·6	76·6

Madang, New Guinea (5°S.; 145° 45′E.)

Temp.	81	81	81	82	82	81	81	81	81	82	82	82	82
Precip.	12·5	11·6	15·0	17·2	14·8	10·7	7·9	4·8	5·8	10·2	13·3	14·41	38·2

4. Topography and Soils

With flatter land at a premium there has been a great deal of reuse of previous gardens. Soils are typical tropical laterites and upland podsols, and react strongly when exposed to heavy rainfalls. Constant burning of forest and later of the kunai and razor grass and scrub regrowth has assisted in depleting soil fertility. Leaching and eluviation is speeded up by clearing and burning, and two decades or more may elapse before the gardens are reused.

5. Cultivation Practice

a. Selection of Site

Few areas today contain any worthwhile virgin forest land. This is especially so where population densities are higher. The Chimbus usually select an area of reasonable slope, which may have been under bush fallow for a number of years. The flatter tops of ridges usually become the village site, and hillside slopes of varying degrees of steepness become the sweet potato patches.

Large trees are felled, using steel axes today, and the saplings, scrub and grass are fired during a drier period. Often the area may be located at a considerable distance from the dwellings, and may be separated from other garden patches by land too difficult to clear or too steep to be worthwhile for agriculture.

b. Preparing the Gardens

After clearing and burning, the site is fenced off with sharpened stakes. This is a necessity to keep the scavenging pigs from the vegetable patch. The simple division of labour typical of subsistence agriculturalists means that men clear while women plant, tend and harvest the crops.

c. Crops

The basic vegetable is kau kau (the sweet potato), together with yams, taro, cassava, beans, coconuts, paw-paw, sugar cane and the dwarf sugar termed pit-pit.

Where European influence is considerable several other crops—cucurbits like pumpkins and melons, and a variety of salad vegetables—are included. The period between cultivation and bush fallow is considerable and the Chimbu

A typical village location of the Subsistence Agriculturalists in the Central Highlands of New Guinea. The village is located on a flattened ridge summit. The gardens are found at varying distances from the settlement on the sloping sides of the ridges. The influence of Western missionaries is evident in the large foreground church and missionary hut in the right background.

Women cultivate garden plots at Lufa, New Guinea. A similar plot can be seen in the right middle ground, and yet another in the left middle ground. Fences of pointed stakes protect the gardens from pigs which forage in the land which is not cultivated.

Clearing a site and cutting fence stakes for a New Guinea Kau Kau patch. The trees are being barked and the 8-10 feet stakes are split from the green wood. The area has been under bush fallow for several years, to the limits of the tree line in the background. After the fence stakes are sharpened, the points are fire-hardened before being driven into the ground (middle distance centre).

An aerial view of Lake Murray, 185 miles north of Daru in the Western District of Papua, and a five-day trip by boat up the Fly River from Daru. The lake area has a population of approximately 1,200 Papuans, who gain a subsistence livelihood by fishing in the lake, hunting, cultivating small food gardens and making sago from wild sago stands. The lake extends over an area of approximately 200 square miles, and the water level rises from 20 to 30 feet during the rainy season.

peoples are at the most efficient end of the scale. Other peoples in areas where soils are less suitable may use the soil for only one crop and leave a fallow period for many years.

The staple sweet potato is grown from cuttings which are stuck in the soil at any time of the year. The tubers are dug as needed and the gardens have plants in various stages of development all year round.

Although the diet of the natives is still based largely on kau kau, other vegetables include yams—particularly in the Sepik-Ramu river valleys and the Trobriand Islands. There are dozens of varieties of taro grown. Sago palm is very important, especially in the lower valley levels and along the coastal plains. The coconut too is very important in these areas, and has become the staple cash crop where demand from the Administration has grown.

d. Chimbu Agriculture

In these areas of high population densities, pure shifting cultivation is not so obvious. The Chimbu peoples, for example, usually adopt a semi-shifting type, with a long-term occupation of village sites. Houses are individual and separated by considerable distance rather than grouped in true villages. Rotation of fields in most areas is short term with one or two years of cultivation and twice this term under bush fallow. Constant burning of the secondary regrowth greatly restricts the length of time the areas can be gardened.

Chimbu gardens stoutly fenced off from pigs represent a high degree of hard work. The fences, ditches and gardens themselves are straight in line and remarkable for their neatness. This is unusual in comparison with primitive gardens in other parts of the world. These Highlanders by their ingenuity have freed themselves from the problems of food shortages, again unusual compared with many other subsistence agriculturalists.

Tools are primitive. Even today many stone axes are still to be seen, while the digging stick is almost universal in many areas. Steel axes and hoes are replacing traditional tools when food surpluses allow trading to be made or when some cash or barter transaction is carried out.

The Chimbu peoples are perhaps unique in their attempt to maintain soil fertility and check soil erosion. They plant the feathery leaved quick-growing Casuarina trees in their garden plots about 12–18 months prior to deserting them and there is a resultant slight increase in fertility when re-use is necessary.

Slash and burn techniques, however, are employed throughout the land, and the ensuing depletion of soil fertility is everywhere obvious. Although pigs are let into overgrown gardens to forage and thus indirectly, fertilise the soil, they are herded for slaughter by burning the field and the soil is left more vulnerable to leaching and water erosion.

6. Other Features of Subsistence Life

Hunting, fishing and gathering are still very important in varying the vegetable diet. The chief animals hunted include species of wallaby, the tree kangaroo, the cuscus, several species of phalanger, bandicoots, ant-eaters, bats, mice and rats. Many varieties of birds are hunted for food and for ornament—particularly the birds of paradise. Freshwater fish are a valuable source of food, and as in most Asian countries they supply much needed protein.

Around the villages the pig is most important, although the pork is consumed in vast amounts during feasts and celebrations rather than as a regular feature of the diet. Dogs and cats are also eaten. Cooking of all foodstuffs is in heated stone ovens or in hollowed wooden cylinders or bamboo tubes. The sections of bamboo are frequently the only utensils used in cooking, and are employed in carrying drinking water too.

The women make net bags to carry produce from the gardens, while clothing is of beaten bark, grass or palm leaves. Head-dresses of feathers and collars and nose pieces of shell are worn by the males. Shell is traded from the coastal areas. Chief exchange is of gold-lip and black-lip pearl shell, clam and cowrie shells and salt from coastal tribes, for yams, taro, sweet potatoes, implements and weapons and bird of paradise plumage.

As Administration officers and missionaries make increased contacts, the subsistence way of life changes into subsistence plus some cash cropping. Steel tools and household equipment make life easier. New plants and seeds are introduced. Coffee, cacao and coconuts are staple cash crops. Improved pig and poultry breeding with introduced stock is a further feature.

Traditional subsistence methods are still the basis of life for most of these peoples, although cash crops for export are increasingly important. In the readily accessible coastal plains, and where airstrips have been constructed, the traditional methods are now largely absent. Native co-operatives, smallholdings and plantations are characteristic here.

From the coast to the inland, gardens are the mainstay of life and centre of magic and tradition in the Trobriand Islands, 100 miles east of New Guinea. The gardening year begins with a conference summoned by the chief and held in front of the magician's house, to decide where the gardens will be made, who will cultivate particular plots and when work will begin. Distribution of the yam crop is complicated and governed by many rules and taboos. The gardener who has grown the crop sends a large part of it to a specific relative, usually the husband of his sister. The carriers stack the offering —in this instance sweet potatoes—in front of the new owner's house.

On a flattened ridge in the Central Highlands of New Guinea, taro (variety Xanthosoma) and young bananas are cultivated.

Bibliography

Daryll Forde, C., *Habitat, Economy, and Society*, Methuen, London 1963.

Essai, B., *Papua and New Guinea: A Contemporary Survey*, Oxford University Press, Melbourne 1961.

Gourou, P., *The Tropical World* (3rd. ed.), Longmans, London, 1959.

Gourou, P., The quality of Land Use of Tropical Agriculture, in *Man's Role in Changing the Face of the Earth*, ed. W. L. Thomas, U. of Chicago Press 1962.

Howlett, D., *A Geography of Papua and New Guinea*, Nelson, Melbourne 1967.

Laut, P., *Agricultural Geography:* Vol. 1, *Systems, Subsistence and Plantation Agriculture*, Nelson, Melbourne 1968.

Questions

1. What is the principal location of primitive subsistence agriculture in New Guinea ? What are the particular reasons for this location ?
2. What are the characteristics of the environment in which this type of agriculture is practised ?
3. In what ways do the Chimbus differ from other people practising this type of agriculture ?
4. Why do you think that other peoples practising primitive subsistence agriculture have not endeavoured to implement such methods as are used by the Chimbus ?
5. Is the practice of subsistence agriculture in New Guinea likely to be changed as the island moves towards independence and eventual economic viability ?
6. Does this type of agriculture have a vital part to play in the role of Papua-New Guinea as an emergent nation ? Discuss fully.

5 Intensive, Rice-Dominant Subsistence Agriculture in Japan

1. Introduction

Acute land shortage in Japan has forced the nation into an intensity of land use which has few comparisons in world agriculture. Of Japan's 100,000,000 people, nearly half are agriculturalists. While the spectacular increase in industrialisation since the Second World War has permitted large scale imports of foodstuffs—cereals in particular—Japan has continued in attempts to lift domestic food production. Despite these attempts, often highly successful, a great proportion of Japan's farmers can still be classed as subsistence agriculturalists.

Rice remains the basic crop—occupying 56 per cent of the total cultivated land in Japan. This crop is dominant in some areas—southern Honshu, for example—to the extent that it occupies up to 85 per cent of the total cultivated land. It is eminently suited to the flat coastal and river alluviums where water is abundant and sub-tropical climatic conditions give an all year growing period. Here two crops per year are possible. In combination with padi fish farming the output per acre is very high. Coupled with this is a high nutritional and calorific value, and long-term storage qualities. Thus the staple rice-based nature of Japan's diet is established.

Where only one crop of rice is possible, the winter cereals—wheat and barley in particular—take over after the rice harvest. Considerable amounts of vegetables and some industrial crops are grown. On land too steep or impossible to irrigate, commercial crops—tea, mulberries (for silkworm feeding) and fruit crops—are common. Yet basically agricultural production is geared to rice. Synonymous with rice is water.

Water holds the key to the cultivation of wet rice. Water not only provides the environment for the growing of the rice, but retards weed growth and makes possible the existence of nitrogenous bacteria which flourish in the water covering the fields and which help refertilise the fields.

Wet growing conditions minimise crop rotation, but maximum production is only reached by the expenditure of vast amounts of human energy. Padi technology is very complex and demanding. For example, soil has to be prepared, fields have to be made to contain water at the exact height required and the irrigation facilities must allow for both flooding of the padi fields and drainage of rain and ground water when the fields are dried. A co-operative community enterprise is required to maintain irrigation facilities. The characteristics of rice-padi agriculture highlight the enormous amounts of human labour required in this intensive type of farming.

2. Location

Japan has 15–16 per cent (approximately 15,000,000 acres) of its total area, largely coastal and fluvial plains, suitable for agricultural purposes. While rice is grown in almost the full length of Japan, the only exception being northern Hokkaido, the

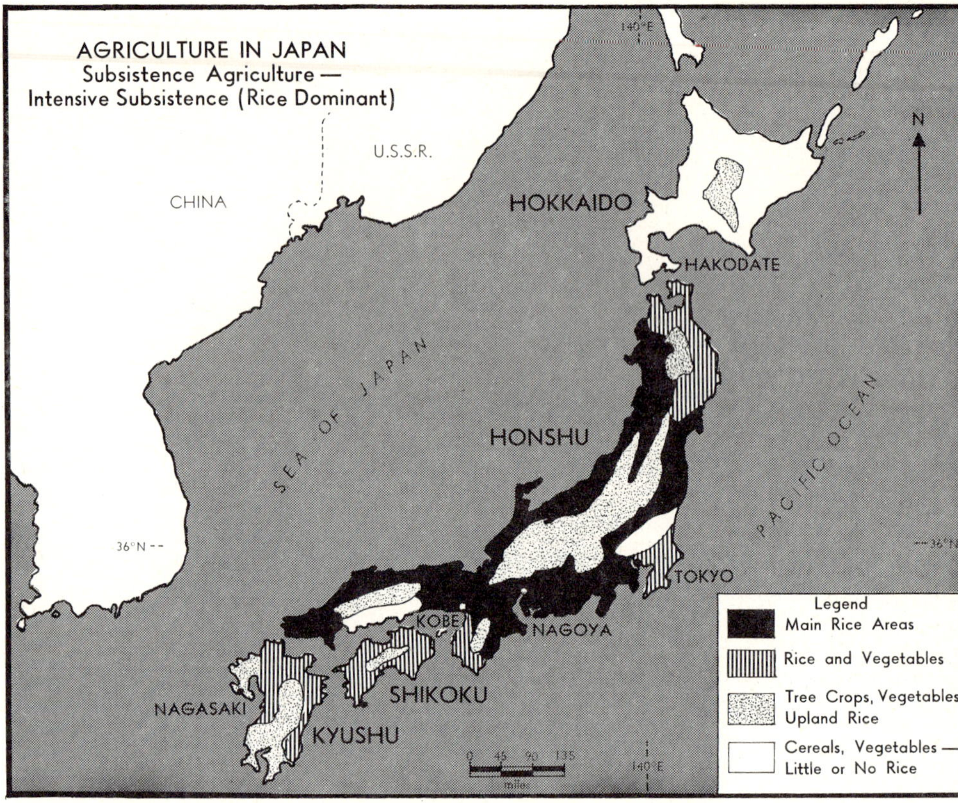

main producing areas are northern Kyushu, southern Honshu and Shikoku. These plains are also the areas of largest urban population and cultivation. The population density of cultivated land is approximately 4,220 per square mile, which is twice that of Belgium's cultivated land. Perhaps a larger percentage of surplus food than elsewhere in Asia goes to feed the inhabitants of the great cities. Such cities are Tokyo, Yokohama, Osaka and Kobe, to name some of the best known.

3. Climatic Conditions

The most favourable conditions for the growing of sawah rice are heavy rainfall with maximum precipitation in summer and also high summer temperatures. Irrigation in hot areas satisfies these conditions. Usually rainfall is supplemented by irrigation water. The universal availability of electric power for pumping water is a great advantage in Japan.

The following statistics illustrate the climatic conditions suitable for rice growing in Japan:

Selected Statistics (temperatures in degrees F.; precipitation in inches)

Tokyo, Japan (35° 42′N; 139° 40′E.)

	J	F	M	A	M	J	J	A	S	O	N	D	Year
Temp.	39	39	45	55	62	70	76	79	73	61	51	42	58
Precip.	2·2	2·6	4·4	5·2	6·0	6·4	5·5	6·4	8·9	7·5	4·1	2·2	61·4

Kagoshima, Japan (31° 48′N; 130° 35′E.)

	J	F	M	A	M	J	J	A	S	O	N	D	Year
Temp.	45	45	51	60	66	72	78	80	75	66	57	48	62
Precip.	3·4	4·0	6·4	8·7	8·2	17·0	12·2	7·4	8·7	5·2	3·7	3·4	88·3

This photograph shows the increasing use of machinery in Japanese rice farming. It is significant that the implements are manufactured in Japan. Even with the aid of machines, note the labour—intensive nature of farming, particularly women labourers.

Japanese rice fields after harvest. This photo clearly shows the intensive nature of this type of farming whereby the small farms are orderly arranged and no land is wasted. In addition, it shows that the Japanese pattern of settlement in the form of individual detached cottages contrasts sharply with other Asian countries practising this form of agriculture.

The type of surface and the soil is of great importance also. Level land is necessary for the padi fields to contain water, and the sub soil must be non-porous so that the water does not escape. The great bulk of rice is grown under irrigation and the rice fields are submerged under about six inches of slowly moving water for seventy-five days at a minimum.

Evaporation rate and soil conditions demand that between 45″ and 65″ of rainfall occurs. Rice also needs an average 70°F. during the four to six months growing season, with ample sunlight intensity in the ripening period.

In northern Hokkaido temperatures are insufficiently high to allow the cultivation and harvesting of rice. However, over most of the suitable areas of Honshu, Shikoku and Kyushu rice is cultivated. Only in Kyushu and Shikoku is the growing of two successive rice crops possible. However, in large parts of southern Japan's agricultural regions, rice plus one other crop is quite common; this crop is often barley or wheat.

4. Characteristics of Rice Growing in Japan

a. Nursery grown seedlings

Intensive culture of seedlings allows time for winter crops to be harvested and an extra period for the padis to be prepared. The use of seed beds means that less land

Increasingly in Japan, cash cropping or even pure commercial farming involving the most modern techniques—in this case, the application of insecticide by helicopter—have arisen from advances in intensive subsistence farming.

Nursery bed for dry field rice farming. Virtually every plant receives individual attention—typical of the meticulous cultivation involved in intensive farming.

is used in the early stages and the smaller areas make it possible to provide much more careful attention to weeding and thinning out where necessary. Water is also conserved and labour is saved. An additional point in the favour of seed beds is that there is a slight mutilation of the plant on transplanting which gives an added stimulus to growth. The rice is planted in North-South rows to give maximum sunlight, which helps give highest yields in the quickest time.

b. Interculture

The growing of other crops between the padis is widely practised in southern areas. The banks separating individual bays within the padis are planted with additional crops of peas, beans, barley, rape or wheat. On a large percentage of the farms, where padi fields are drained, winter crops are also cultivated. Intertillage is carried on in the most favoured areas and here it is possible to grow three vegetable crops in the one year. This occurs particularly in the better areas near the big cities.

c. Multiple cropping

Multiple cropping is possible in the most suitable areas, which are entirely below 37°N. Latitude. An extra crop of rice is planted after the first crop is harvested. Transplanting is essential where two crops are grown each year. This is laborious, as it has to be done by hand. However, it saves time because growth begins in the seedbed, thereby allowing the use of the padi for the maturing of another crop or for late preparation for wet rice.

It has been estimated that the transplanting technique saves several weeks and thereby makes possible the growing of two crops on the same land. In exceptionally favoured parts of Japan, three crops are possible by the extensive use of fertiliser. Improvements to padi drainage and the consolidation of fields will allow multiple cropping to be increased.

d. Terracing

Terracing is an important characteristic of wet-rice cultivation in Japan and is indicative of an agricultural land use which is at the same time land-intensive and labour-intensive. It is doubtful now whether the construction of terraces is an economic proposition. Only the smallest machines could ever be used in terraced areas and as a result yields are hardly sufficient to cover capital outlay.

The draining of swamps and the building of terraces have made possible a slight increase in cultivated areas. However, this procedure cannot be continued indefinitely because of the lack of suitable areas and the great cost involved. Agrarian reform after the Second World War led to the disappearance of absentee landlords and the beginning of the farmer operating his own small plot and being able to pour back into his farm any surplus profits which he had available. Increased individual attention to all aspects of farming resulted in an increase in rice production. Yields have also been increased by the more ready use of artificial fertiliser and the development of improved types of rice seed.

e. Use of fertilisers

The Japanese rice farmer uses all types of fertiliser—organic fertilisers such as manure, night soil and rotting leaves and also inorganic fertilisers such as ammonium sulphate, superphosphate and potash. Approximately 1,000,000 tons of phosphate is imported into Japan each year to swell the amounts of artificial fertiliser used on the farms. Limestone is also used to minimise the acidity of soils in certain areas.

Dobby estimates that Japan consumes 78 pounds of fertiliser on a per capita basis, compared with Taiwan (39 pounds), Ceylon (12 pounds), and India (1 pound). If farm area is taken into account the fertiliser used can be estimated at 325 pounds in Japan, 168 pounds in Taiwan and 41 pounds in Ceylon. Japan's yields are correspondingly high, as can be seen from the following table:

Japan	79 bushels per acre
China	55 bushels per acre
United States	50 bushels per acre
India	23 bushels per acre

If we exclude small producers such as Italy and Spain, then we can say that Japan has the highest yield per acre in the world.

The use of fertiliser in Japan has reached the stage where it is doubtful if increasing usage is an economic proposition. Research is being carried out to investigate the cost of increased productivity through increasing use of chemical fertiliser.

An intricate pattern of heavily terraced hillsides in central Japan. Wherever possible, the land has been banked to provide additional rice padis. Such terracing is characteristic of large parts of South-East Asia's rice lands.

f. Seed stations

Recent developments of seed stations have been important in helping the production of rice. Seed stations scattered systematically throughout the country provide the best grain (the result of scientific seed development) for planting in each area.

g. Mechanisation

In recent years mechanisation has progressed to the extent where an estimated one in four farmers uses the two-wheeled tractor for ploughing. Machines are also used for threshing, drainage, watering and chemical spraying. Except in Hokkaido, sowing, planting, and harvesting are not mechanised.

5. The Calendar of a Typical Rice Farm in Southern Honshu

Although other crops are produced, rice is the staple food, the basic crop. Ninety per cent is ordinary rice, the rest glutinous.

1 May The seedbeds are prepared (this follows winter fallow of certain small areas). A central location is chosen so as to facilitate planting out at a later date. At this time the main fields are still occupied by dry winter crops.

The seedlings have to be available when the dry winter grains are removed.

The seedbeds are ploughed and irrigated, and the soils prepared by working with hands and by using fertiliser. Infertile grains are floated off from the fertile grains in salt water.

The surface of the seedbeds is prepared very carefully and divided into rectangular

The main irrigation canal (over 100 miles long) of the Aichi Reservoir project in central Japan, south-east of Nagoya city. The canal is crossing the heavily settled Lobi Plain, where rice and a wide variety of vegetable crops occupy much of the area. This area typifies the endeavours of the Japanese to extend irrigation farming by modern scientific methods into areas previously impossible to farm intensively.

beds and the rice is broadcast. Scarecrows are erected to frighten the birds away.
Mid May The seedlings begin to appear.
Rice borer and other pests are dealt with by spraying with insecticides such as DDT and BHC.
Late May Winter crops of wheat and barley are harvested.
1 June onwards The main padi fields are prepared for transplanting. Compost and chemical fertiliser are used, the soil is ploughed twice, irrigation water is turned on, and the soil is harrowed and smoothed.
20 June onwards By this time the seedlings (8–10″ high) are ready.
They are taken to the padi fields where groups of three are planted as a single plant. All family help is required at this time, even school children and people who work in city offices if possible.
The transplanting out is laborious work and great care must be taken with the seedlings. They are planted about eight inches apart in perfect rows 1½–2 feet apart.
July–August The rice padis are weeded three times with the assistance of a mechanical cultivator.
Chemical fertiliser is applied twice in this period. The usual types are ammonium sulphate, lime sulphate and potash.
End of August Insecticide is again sprayed to get rid of rice borer and leaf hopper.
September Weeding is done by hand.
Mid October Padi fields are drained.
Much labour is needed to cut, dry, thresh, winnow and hull the new rice.
When the rice is dried it is threshed.
November While the rice is drying out, the land is prepared for winter crops. Barley and wheat are popular crops in this period.

6. Other Crops in the Rice Dominant Pattern

Although it has been indicated that rice is dominant—in warmer Japan at any rate—the growing of other crops is characteristic wherever conditions permit. These other crops are usually grown on steep slopes which may be partially terraced. Poor sandy soil with a clay base is made suitable by liberal quantities of night soil, which is added constantly throughout the year. Dry field multi-cropping is carried on where possible all the year. In southern Japan the growing of three dry field crops per year is common practice and all fields have at least two crops.

Early summer is the usual time to plant barley, rape, broadbeans, peas, giant white radish, onions, carrots, cucumbers, spinach, cabbage and lettuce. Late summer is the period for soybeans, red beans, cowpeas, sweet potato, sesame, carrots and water-melon. During winter cabbage, spinach, edible chrysanthemums, leeks and the giant white radish are predominant.

The sweet potato, even though it constitutes only 6 per cent of the food consumed, is very important because it is available for consumption in winter and late spring when the grain supply is most depleted. The giant white radish is also important because it can be planted in both summer and winter and provides valuable supplementary food for the predominantly subsistence farmers.

7. Trends in Rice Production, and Conclusions

The Japanese rice farmer is much more advanced and scientific in his methods than other rice growers in Asia. The impact of government agricultural stations is widespread. Advice and assistance in all aspects of cultivation is freely available, and the research workers have made important break-throughs in techniques.

Select, superior strains of seeds—the result of careful breeding—have raised yields. The introduction of chemical weedicides and pesticides has helped to combat partial crop failures or low yields. Chemical fertilizers have received great attention in recent years.

The farmer himself has several advantages. The fact that the bulk of the population are literate enables the dissemination of the latest trends through farm bulletins and journals. Spectacular development in communications, in particular radio and television, has allowed information on techniques, market reports and the like to reach huge numbers of peoples.

It is doubtful if the combination of the great advances of recent years will permit the transition to commercial agriculture. At best the subsistence farmers will be reduced in numbers, and collective techniques will permit some increases in cash cropping.

In a country where the average agricultural holdings are only 2¼ acres, and one-third are 1¼ acres or less, the high yields are a tribute to the ingenuity of the farmers. At the same time the above figures highlight the acute shortage of agricultural land and indicate a continuing reliance on grain from foreign countries in the future.

Bibliography

Alexander, J. W., *Economic Geography*, Prentice-Hall, Englewood Cliffs, N.J., 1963.

Beardsley, R. K., and others, *Village Japan*, University of Chicago Press, Chicago 1959.

Ginsburg, N., *The Pattern of Asia*, Prentice-Hall, Englewood Cliffs, N.J., 1963.

Grist, D. H., *Rice*, Longmans, London 1960.

Matsuo, T., *Rice Culture in Japan*, Yokendo, Tokyo 1961.

Statistical Handbook of Japan, Bureau of Statistics, Office of the Prime Minister, Japan.

Trewartha, G. T., *Japan*, Methuen, London 1945.

Questions

1. 'Most Asian rice areas are now aiming for higher levels of agricultural efficiency and living standards. This largely means new rice technology. In this context, Japan can be considered in the forefront of such development.' Write an essay based on the above statement.

2. In what ways is the progress in methods of Japanese rice farming related to her advance in industrial techniques?

3. In what ways do the Japanese rice farmers' techniques differ from those adopted by most other Asian rice farmers?

4. What significance and relevance has the Japanese progress in methods of cultivation in FAO programmes?

5. 'The Japanese economy shows marked commercial facets (even in agriculture) yet its rice farming still largely reflects long-standing intensive sedentary and subsistence features.' Explain and account for the above statement.

6. 'The interplay of cultural and natural factors has profoundly influenced the production of rice and indeed all crop combinations in the Asian countries.' Critically discuss the above statement with particular reference to Japan.

Part Three

Plantation Agriculture

6 Plantation Agriculture

1. Introduction

'Plantation Agriculture' has consistently posed many problems of definition. This is largely due to the traditional links between plantations and particular products, such as rubber, sugar, cotton, coffee and others. These are frequently, but loosely, termed plantation crops. Yet significant amounts of these products are grown by individual cultivators or 'smallholders'. It is important that the link with a particular crop be broken, otherwise errors perpetrated by Whittlesey (and others) in classifying the Queensland 'sugar coast' as an area of Plantation Agriculture, for example, will persist.

2. Basic Considerations

Any attempt at defining Plantation Agriculture will include a discussion of the following points.

i. The Crop or Crops Concerned Are Tropical Or Sub-tropical In Type
Two points are raised here—monoculture or a crop association of two or three types; and physical location.

a. Monoculture or crop combinations. While it is often usual to think of a single crop as being typical, this is by no means always the case. Gregor cites Californian examples, and this is also true of much of America's Deep South, where a wide range of crops may be grown in combinations usually of two or three of the following—cotton, rice, sugar, soy beans and pecan nuts. Often included with these are beef cattle and hogs. In other parts of the world there is frequently one crop which is dominant—cocoa, palm oil, rubber, copra, coffee, for example. The practice of monoculture has the advantage of creating economies of scale, although it carries with it the inherent danger of specialisation in projects which are often subject to depression in periods of overproduction. In general terms Plantation Agriculture can, however, be considered monocultural. The bulk of the world's output is produced under this system.

b. Location. Geographers have frequently attempted to categorise agricultural activities under broad climatic headings such as Equatorial, Tropical and Temperate —or an extreme case as in 'Mediterranean' agriculture. (The danger of classifying wheat as a 'temperate' crop, for example, can be gauged from its distribution within the tropics in India to practically within the Arctic Circle in Siberia.) Because of the wide range of crops grown, plantation agriculture is especially susceptible to the generalisation of Tropical or Sub-tropical, yet there is no way around it. Plantations

THE WORLD
SHOWING AREAS OF
PLANTATION AND ESTATE AGRICULTURE

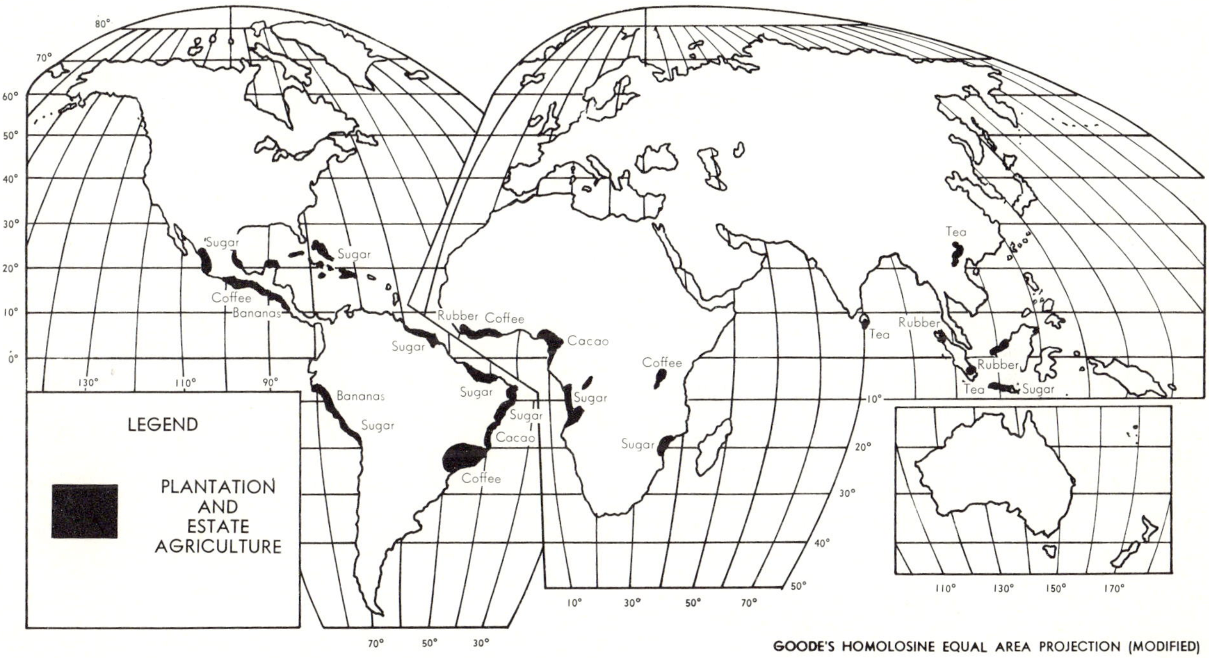

are found over a wide range of latitudes from the cotton fields of America's south to the Equatorial rubber groves of Indonesia.

ii. The Two Plantation Types

Frequently large scale plantations and native smallholdings exist alongside one another, yet there are significant differences of degree:

a. The large scale enterprise. The Estate type plantation is the type commonly studied and cited in most texts. In size it is considerably larger than a 'family' farm. It contains a community of workers and their families, a headquarters with offices for administration and a processing plant in which the crop is prepared for transport overseas. The Estate was established by European capital and enterprise with the intention of producing export-oriented commercial crops for consumption in temperate lands. The Estate has been organised by the use of indigenous labour or by immigrants from areas where a cheap work force was available.

The administration and technical staffs were European. In recent years greater opportunity has been given the local people to take more important and more highly trained positions. In lands where plantations have been nationalised and where colonial policy had made little provision for technical training of the indigenes, inexperience has frequently created a breakdown in administration.

In sugar plantations, such as in Hawaii, there are large irrigation installations, tractor-powered planters, machines for cutting, piling and loading, and to operate these a considerable trained staff is needed, both in the field and in the office. In Malaya's rubber plantations many people are needed for tapping the trees and for the preparation of latex for export. On Central American banana plantations considerable labour is required for planting and year-round cutting. Technicians are needed to control sigatoka by spraying oil fungicide (often from helicopters) on to the rows of bananas, and also to keep the crop clear of Panama disease. Costly machinery is necessary for the processing of such products as sisal, sugar cane and palm oil.

Plantations have extensive buildings; for example, the coffee *fazenda* in Brazil has large storage sheds and drying facilities, a laboratory for the scientific investigation of diseases and new types of plants, a hospital and houses for the *fazendeiro*, his foremen and the labourers and their families. The coffee berries have to be dehulled and washed and the beans dried in the sun. Few large plantations nowadays fail to have facilities for processing even though it may only be the initial stage. On cacao plantations for example, the pods are opened and the beans removed to increase the value per unit weight prior to transportation to the dryer.

These points stress that the plantation may be a large scale enterprise enjoying economies of scale, yet, as indicated earlier, significant amounts of plantation products originate from individual farmers.

b. Native smallholdings. One of the basic considerations which differentiate Estates and Smallholdings may be the crop type. Figures for the respective share of native and Estate holdings in Indonesia have revealed this clearly:

Table 1 Respective Share of Peasant and Plantation Agriculture—Some Selected Crops

Commodity	*Peasant Agriculture %*	*Plantation Agriculture %*
Palm oil and Kernel	Nil	100
Sugar	1	99
Tea	18	82
Rubber	48	52
Kapok	83	17
Pepper	99	1

It is obvious that crops requiring large scale investment, and complex and costly machinery are best suited to plantation systems. The higher the percentage of hand labour employed, the greater the percentage which will be produced by individual growers. This ratio is never fixed, and may change over time with respect to a single crop. Sugar for example could be cited as one of the most complex of plantation products, yet even here the mill owner can co-operate with peasant producers who sell the raw cane. There are areal variations too—cacao in West Africa is basically a peasant crop while in the tropical regions of America and Asia large amounts are produced by plantations. Sometimes governmental policy may effect a change, in both Malaya and Indonesia official policy has favoured plantations at the expense of peasant holdings.

Thus the size of the native 'plantation' may in fact be quite small. Cocoa production in West Africa operates on a contract basis through the Unilever combine; in newer areas of Brazil individuals grow cacao on areas less than fifty acres; an extreme case occurs in the Bas Dahomey region of coastal Guinea where areas are from three to twenty acres and portion only is given over to crops like palm oil, in association with root crops and vegetables.

Table 2 Some Comparisons of Plantation ('estate' type) and Peasant Smallholding Agriculture

Plantation (estate)	*Peasant Smallholding*
1. Uniformity of crop and product	1. Often several varieties
2. Ease of control of disease	2. Hand methods of doubtful value
3. Large scale planting of improved varieties	3. Individual cannot finance research
4. Efficient production claimed	4. Levels of efficiency lower
5. Social problems of contract labour (i.e. management-staff relations)	5. Private ownership incentive
6. Large capital investment; inflexible in time of market glut	6. Easier to restrict or change emphasis on products

iii. Export Orientation

Plantations specialise in export-oriented cash crops grown in tropical and subtropical areas which find their markets largely in the technologically advanced countries of the mid-latitudes. These countries consume vast quantities of all types of plantation products. These products range from Fijian sugar and bananas produced in the Azores to West African palm oil and cacao from Ecuador.

The obvious reasons for this market orientation are the tropical environment and nineteenth century colonisation. European traders and later settlers have for centuries been transporting the foodstuffs and industrial crops to urban processing areas in their own countries. Originally these crops were the result of gathering rather than agriculture. The rise of Western cities and Western markets and their expansion demanded considerably increased quantities, and crops were farmed in suitable locations, or, as in the classic case of rubber, were re-located. Europe, and later the United States, supplied a great percentage of investment capital, machinery, managerial staff, soil and plant scientists, fertilisers and pesticides and transport facilities. It was quite natural under these conditions that they should process and market the finished products.

This position largely remains today, with the proviso that there are areas of change and expectation of much more change as nationalism continues to increase.

iv. Labour

The section above introduces entrepreneurial control, which is but one aspect of labour force. On both estates and in areas of smallholding production the foreign-born, foreign-trained manager is still very common. Since the Second World War many countries have been granted or have seized independence and control has passed into the hands of the indigenous peoples.

There have been notable problem areas. Indonesia and Central Africa have in particular witnessed problems of internal management and export marketing. Cuba's political revolution has placed sugar production under nationalisation and this, together with the loss of the American export market, has created severe economic problems. Many of the problems have their bases in lack of managerial skills when American exporters were relieved of their duties. Similarly, Indonesia suffered an economic breakdown in her plantation industries when Dutch and British plantations were nationalised in 1958 and 1964 respectively. In few agricultural activities is skilled know-how more important than in plantations.

The bulk of the work force on estates is unskilled or semiskilled. This reflects two important aspects of plantations—the high percentage of hand labour employed, and the simple initial processes carried out on the plantation site.

Historically, plantations have drawn their work force from the indigenes where practicable; slave labour in the eighteenth century and immigrants when others sources failed; for example, Japanese coolies in Hawaii, Indians in Malaya's rubber estates and in Fiji's banana and sugar plantations.

Portion of the large capital investment in plantations must be allotted to housing and service needs for the work force. Wage rates are not high, especially since the bulk of the work is routine, and there is usually constant supervision by the administrative and technical staff. Yet despite the pool of cheap semi-skilled labour, the system introduces serious social problems unless care is exercised. In the choosing of a site for the estate the plantation authorities have to be careful to avoid native food-producing areas in order to obviate the possibility of disturbing the social framework. Also, these areas would be available only at a premium.

It is also essential that nothing be done to prejudice the availability of the pool of native labour. When Henry Ford acquired 2,500,000 acres of land in the Amazon Basin to establish his Fordlandia and Belterra plantations the natives were paid too highly. This resulted in the disruption of the local labour market, antagonised other employers, and reduced the amount of work performed considerably. The reason was that the native labourers by doing half the work done previously could earn just as much money. Two days' work each week was enough to supply their meagre wants. In order to get satisfactory work performed it was necessary to reduce the wages to 60 cents a day—just slightly above the amount paid by other employers in the area.

v. Perennial or Annual Crops

Perennial trees, such as rubber, are obvious examples of crops which incur high capital establishment costs and a waiting period when costs are high for no cash return. Rubber, depending on soil conditions and agricultural practice, may be continually tapped for periods of 50 to 100 years, and most other trees, like cacao and coffee, from 10 to 20 years. This raises the problem of capital tie-up, heightened in periods of falling market prices. The inflexibility is also obvious in terms of output which can not be quickly expanded or contracted.

Against this is the corollary that once the trees are planted no further costs are necessary for long periods. The longevity of the trees works to the plantation's advantage, and production continues for many years. Thus considerable savings can be incurred by careful planning of labour utilisation. Also, long term contracts for a portion or all of the crop bring some price stability. Too often, however, there are wild fluctuations in prices which depress the industry—Malayan rubber and Brazilian coffee are obvious examples.

In the face of the above points there is increasing pressure on the estate system to give way to smallholders, and the smallholders in turn are moving in the direction of increasing diversity of activity.

Annual crops naturally suffer less from the vagaries of market fluctuations. There is greater flexibility in employing capital, in controlling output and especially in diversifying crop types. The restriction on the last aspect is a physical one related to soil types and climate, or an economic one related to market orientation and prices. A changeover to production of annuals may however involve the same problems as in perennial crops, that specialised machinery—as for example cotton pickers—cannot be used for other purposes. In summary the inflexibility is least with annual types.

vi. Special Physical and Economic Facilities Required

a. The physical requirements for plantations. As indicated, plantations are tropical and sub-tropical. The climatic conditions required are relatively high temperatures and abundant rainfall creating an uninterrupted growing season; though some types, notably perennials like cocoa and coffee, require rather specialised ripening conditions.

Selected Statistics (*temperatures in degrees F.; precipitation in inches*)

Rabaul, New Britain (4°S; 152°E.)

	J	F	M	A	M	J	J	A	S	O	N	D	Year
Temp.	82	82	82	85	82	82	81	80	82	82	82	82	81
Precip.	14·8	10·4	10·2	10·0	5·2	3·3	5·4	4·7	3·5	5·1	7·1	10·1	89·8

Rio de Janeiro, Brazil (22°S; 43°W.)

Temp.	79	79	78	75	72	70	69	70	70	72	74	77	74
Precip.	4·9	4·8	5·2	4·3	3·1	2·3	1·7	1·7	2·6	3·2	4·1	5·4	43·3

Kano, Nigeria (12°N; 8°E.)

Temp.	72	77	85	89	88	83	80	78	80	81	78	73	80
Precip.	0·0	0·1	0·1	0·5	2·5	4·8	8·2	12·1	5·5	0·5	0·0	0·0	36·3

Soils are not so important for deep rooted trees like rubber, but coffee in particular prefers a high level of humus in the 'A' horizon. The tropical soils, because of rainforest vegetation and high rainfall usually contain little organic matter. Serious soil deficiencies can only be overcome at great expense and frequently with difficulties of transport when large amounts of fertilisers are required. In the sub-tropics and on small estates the problem is less—rotation, nitrogenous legumes and cultivation methods lessen the difficulties.

Allied with soil and climate requirements are those of topography, particularly slope and drainage. Trees producing tropical fruits and beverages usually require sharp drainage on the lower levels of hills—an obvious erosion risk in view of precipitation amounts. Sugar cane prefers the lowland alluviums and lack of drainage here can be a problem.

Two special circumstances are shelter from sun and wind. Coffee, like many tropical plants, prefers semi-shade, and thus shelter trees must be planted. Bananas grow in the open but require protection from strong winds which will snap mature bunches of fruit, and cacao similarly will drop pods prematurely if not protected.

b. Economic facilities required for plantations. Labour supply has been mentioned as an essential consideration in plantation planning, and thus it places restrictions on certain areas unless migrant labour can be obtained.

Transportation, both into and out of the production zone is hampered by the environment, and thus most plantations require access to rivers or roads and railways where possible.

Closely tied to transportation lines are the world market areas. These are distant markets and transport costs frequently enter into consideration. The stability and long term nature of the market must be considered, and in recent years the increasing competition from synthetics has become apparent. There have been shifts in areas of production to take advantage of newer expanding markets, and there are the slowly emerging markets in countries of origin.

Finally there are special political-economic considerations in many areas of plantation production. Examples have been cited from S.E. Asia—particularly Indonesia, and in Africa. There is serious political instability in many regions which affects production and its costs, and restricts the entry of much needed capital. Political stability in a prospective plantation area is, of course, of paramount importance. No company would be confident of investing in an area which experienced a series of coups and counter coups, as in Sukarno's Indonesia. An environment of political instability is hardly to be sought after by overseas investors. Nigeria which may have appeared attractive to prospective investors in palm oil production previously would certainly not appear so since the tragic dismembering of the country and the breakaway of Biafra. The disruption of the production of coffee in Kenya during the Kikuyu's Mau Mau terrorist movement is a good example of the instability of indigenous peoples bristling under foreign administration. The disruption caused by the breakdown of transport facilities, the breakup of the pool of indigenous labour and the social order in a politically unstable climate is anathema to the operations of the plantation.

3. Some Plantation Crops

While there are many diverse crops grown under plantation systems, they can be grouped as follows:

(i) Tropical fruits, including nuts
(ii) Other foods and beverages
(iii) Industrial and domestic fibres
(iv) Other industrial crops

i. Tropical Fruits, Including Nuts

The major representatives of the group include bananas, pineapples, mangoes, paw paws and coconuts as fresh fruit, and a large variety of nuts.

a. Bananas. Bananas are native to South Asia where they are also grown at present as a subsistence crop. They have spread widely into South-East Asia, Australia and the West Indies.

The 'trees' (actually gigantic plants) require hill-slope drainage and a highly fertile soil, for they are gross feeders and rapidly deplete soil nutrients. In combination they require warm temperature conditions and a high rainfall of 80″ per annum or adequate irrigation. The fruit is soft, and careful handling to markets is essential.

b. Other tropical fruits. Of the other fruits, pineapples are most important, being consumed fresh when marketing is possible and most importantly as a canned product in a variety of ways. Although pineapples are native to Brazil, by far the most important area of production is now the Hawaiian Islands, which produce more than three quarters of the world's trade fruit. Other tropical fruits are similarly marketed, fresh or as additions to canned fruit salad.

All these crops require highly fertile soils with moderate rainfall and warm temperatures. Drainage again is important.

c. Nuts. Much of world production—as in Brazil, for example—is obtained by gathering. The coconut has limited value as fresh fruit but is very important in copra production from which oil is obtained.

ii. Other Foods and Beverages
Sugar, coffee, tea and cocoa are the major representatives of this group.

a. Sugar. This plant, which is indigenous to Melanesia, is now widely distributed in South and South-East Asia, Australia and the Caribbean lands. Although by nature a perennial crop, it is usually set out every two to four years as a ratoon crop that is suckering from the original plant. However, there is a decline in sugar content from the ratoon crops so that two or three years is usually the maximum period before replanting is necessary.

The crop requires around 60″ annual rainfall with a distinct dry season for ripening and moderate to high temperatures throughout the year. The crop does best on highly fertile lowland soils, which need to be relatively flat for ease of transport of the cane and harvesting operations.

b. Coffee and Cocoa. Although coffee is a tropical crop locationally, it does best at high elevations often over 4,000 feet and even up to 6,000 feet. The tree does best on well drained slopes and requires shade trees and fertile humus-enriched soils. High temperatures are required in the growing season with moderate (50–60°F) conditions for ripening. Rainfall must be abundant during the growing period, but drier conditions are necessary for ripening and open air drying of the seeds.

Cocoa, produced from the cacao tree, is also a shade-lover, and a dense overhead cover is necessary. Shelter, drainage and fertile soils are necessary together with a high rainfall—often over 100″ annually—and warm to hot temperature conditions.

c. Tea. This tree grows over a rather wider range of latitudes than either coffee or cacao. It requires a hillside location with abundant rainfall to produce a profuse growth of young leaf 'tips'. Hand labour in picking is a very important factor, and the method of plucking has an important bearing on quality—this being based on the number of leaves taken by pickers.

iii. Industrial and Domestic Crops
Included here are a number of commercial industrial types—the major ones being manila, sisal and jute. Cotton, the chief representative of domestic commercial fibres, is also the major representative of the sub-tropical plantation types.

a. Hemp. The term 'hemp' is commonly applied to abaca (manila), henequin (sisal) and Indian Hemp (jute).

All these crops require a sub-tropical environment with a long warm growing period. Hand labour is used almost exclusively in the production of these crops. The climatic range of these crops is fairly broad; rather lower annual temperatures (50–70°F range) and precipitation around 30″ per annum for sisal and manila, while jute requires much higher rainfall conditions—over 100″ per annum and higher temperatures from 70–100°F.

Although production is found in the south-east and eastern Asian lands it is also widespread in the Soviet Union, Central and South America and eastern Africa.

Jute is mainly concentrated in India-Pakistan, where it does well, even on inundated lands.

b. Cotton. This crop is the most important of the plant fibres and because of this considerable advances have been made in extending the range of the crop climatically. Today the main areas of production are in the United States and Mexico, China, India-Pakistan, Egypt, Brazil and the Soviet Union. Of these, the United States produces over two-thirds of the world's supply.

Cotton requires a growing season of over six months and a low to moderate rainfall around 20″ per annum (or its irrigation equivalent). Warm conditions combined with a drier ripening period in the rainfall pattern are ideal. In the United States, this is one crop in which mechanisation has rapidly replaced hand labour, though the same is not yet true in many other areas. For this reason cotton may not resemble the typical picture represented by some other plantation crops.

c. Rubber. Although Brazil is the home of indigenous rubber, the crop has found similar climatic conditions in several parts of the tropical world, in particular the Malay Peninsula and Indonesia and the surrounding coastal lands where more than 90 per cent of the world's supply is produced.

The crop requires a well-distributed annual rainfall of around 80″ and moderate to high temperatures of 70–80°F. Distribution of rainfall is vital if the flow of latex is to be kept up constantly throughout the year—here rubber is in contrast with some of the other plantation crops, which produce only one crop per annum. Soil type is not vital, for the tree is amenable to a variety of soils, though naturally it prefers fertile areas. Rubber probably more than most crops illustrates the importance of export to world manufacturing centres in Western Europe, North America, Japan and other technologically advanced countries.

Bibliography

Alexander, J. W., *Economic Geography*, Prentice Hall, Englewood Cliffs, N.J. 1963.

Augelli, J. P., 'Bananera: A Tropical Plantation on the Pacific Lowlands of Costa Rica' in *Focus on Geographic Activity* (ed. Norman, R. S., and Patton, D. J.) McGraw Hill, New York 1964, pp. 30–36.

Boesch, H., *A Geography of World Economy*, Van Nostrand Princeton, N.J., 1964.

Fryer, D. W., *World Economic Development*, McGraw Hill, New York: 1965.

Gregor, H. F., *The Changing Plantation*, Annuals Association of American Geographers, Vol. 55, June 1965, pp. 221–38.

Hutchinson, H. W., *Village and Plantation Life in Northeastern Brazil*, University of Washington Press, Seattle 1957.

Langdon White, C., Griffin, P. F., McKnight, T. L., *World Economic Geography*, Wadsworth Publishing Co, Belmont, California; 1964.

Laut, P., *Agricultural Geography;* Vol. 1, *Systems, Subsistence and Plantation Agriculture*, Nelson, Melbourne 1968.

Thoman, R. S., *The Geography of Economic Activity*, McGraw Hill, New York: 1962.

Questions

1. List the distinction between plantation agriculture of the 'Estate' type and smallholdings.
2. What are the physical requirements for tropical plantation agriculture?
3. Plantation agriculture has 'encountered many physical and social obstacles, not all of which have been successfully surmounted'. Discuss.
4. What trend in plantation agriculture might support Gregor's claim that traditional plantation methods are undergoing change? Consider both social and economic developments.
5. 'Plantations have done much to open up numerous tropical areas, through various social and economic factors. Explain their uneven distribution.' Expand upon this statement using at least two examples.
6. 'Plantations are relatively advanced, commercial-cum-industrial systems, located in the tropics. They are large scale enterprises growing, and usually partially processing one or several products. These products are well suited to the tropics, but find their markets in the mid-latitude countries.' Critically analyse this passage.

7 Coffee Plantation Agriculture in Papua-New Guinea

1. Introduction

Coffee was first introduced into New Guinea late in the last century by the German administration. It was not until 1950 that a serious attempt was made at commercial production, when less than thirty tons of beans were exported. The rapid growth of the industry was reflected in the 1965 area and production figures: in this year 261 holdings totalling 12,220 acres (with over 7,800,000 trees) produced 3,374 tons of coffee beans. Today about 93 per cent of the total area under coffee is in the Western and Eastern highlands and the Morobe Districts of New Guinea; although area under production is still increasing, conditions in recent years have at times reflected some of the typical problems of Plantation Agriculture.

World overproduction—particularly in Brazil—has influenced prices, which have fallen steadily since 1957, reaching their lowest level in 1963. In this year and in 1964, Brazilian production was reduced due to severe damage from frost, drought and fire, and she was forced to draw on stocks to maintain her export quota. This gave Papua-New Guinea production a much needed lift. The Territories at the moment have their production classed with Australia, and this avoids an export quota restriction placed on other countries by the 1962 International Coffee Agreement. However, the Territory Administration has ceased to encourage smallholder plantings in the Highlands, and no further areas are being released. Future production for the time being at least will be increased as existing leasehold is put under trees, and as areas already planted mature.

2. The Coffee Plant — Botany and Varieties

All the species of coffee belong to the order Rubiales and the family Rubiaceae, and four species, all of which occur in the Territories (together with many hybrids) are of commercial importance. These are: *Coffee arabica L.* (Arabian or Arabica coffee); *C. robusta L.* (Robusta coffee); *C. Liberica* (Liberian Coffee) and *C. Stenophylla.*

The Robusta and Liberica types are the most important varieties grown at lower altitudes, while Arabica is the chief highland type, and is the most important single type. Quality, too, varies considerably; Robusta is a small bean type of high quality, whereas Liberica is a larger fruiting and poorer quality; Arabica, a smaller bush because it is grown at higher altitudes, is considered to be generally a satisfactory type in quality.

The coffee bushes average around 10–15 feet in height, although they may grow twice this height if left unpruned. They bear clusters of white and tubular shaped flowers at the nodes on new lateral wood only, and this is significant in pruning. Pollination may be effected by wind or insects, although many types are said to be self-compatible and normally self-pollinated.

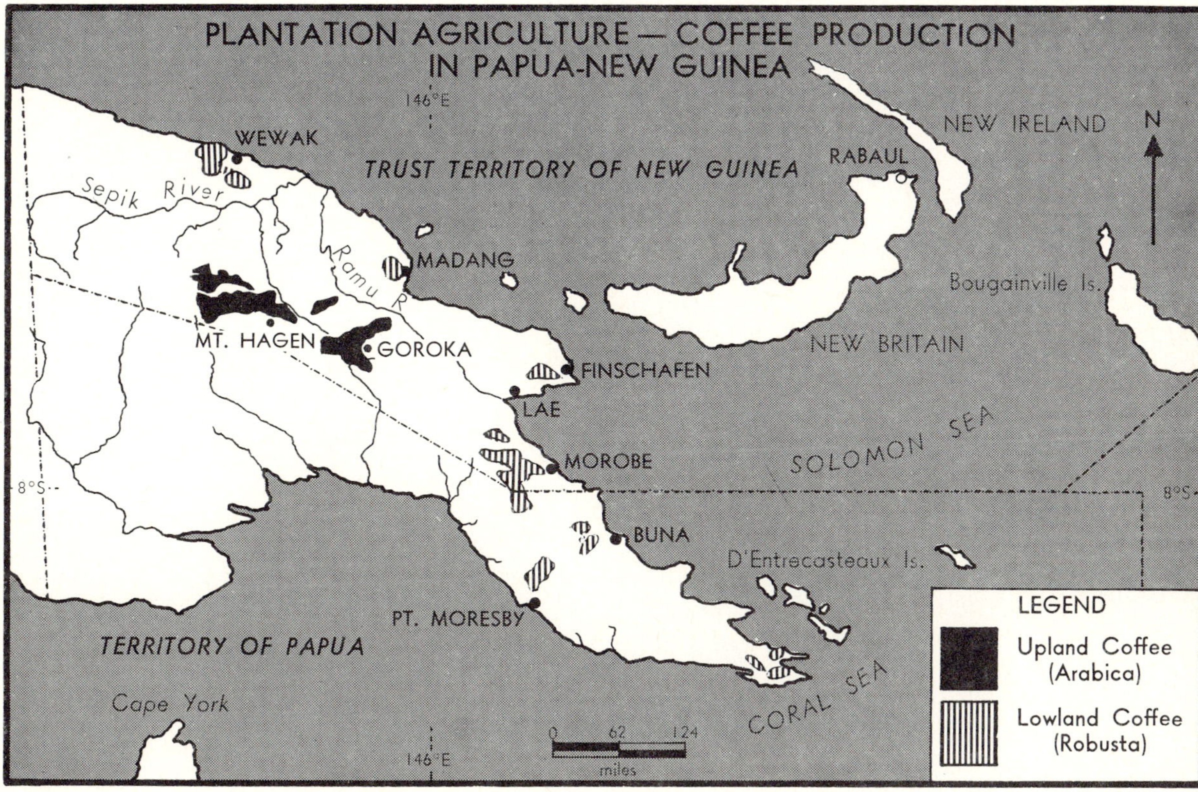

The fruit is a two-seeded drupe resembling a cherry, with one seed in each locus. The sweet pulp is removed by combination of mechanical and fermentative processes, and the hard parchment and underlying silverskin are also removed mechanically to give the coffee bean of commerce. The well known stimulant 'caffeine' is the main alkaloid in coffee, and can be extracted from the waste, and from broken beans. Drug secreting properties are common to all members of the family Rubiaceae.

3. Conditions Required for Plantation Production of Coffee

a. Climatic Conditions and Shade

Allowing for differences in variety, most coffee production is carried out between 1,500 and 6,000 feet altitude in the Territories. Here fairly uniform temperatures ranging from about 60–80°F are experienced. These temperatures must be augmented with a rainfall averaging 50 to 90 inches per year. Distribution of rainfall is important, and less rainfall is desirable during the cooler part of the year when flowers and fruit are forming.

Slope too, is necessary, so that the plants may benefit from good water and air drainage—the latter being vital, if killing frosts are possible.

A marked physiological need is shade for the coffee bushes. This fulfills the dual purpose of sheltering the plants from strong winds, and from sun scald on the ripening fruit. In Papua-New Guinea, large permanent shade trees of the species Albizzia and Grevillea are satisfactory.

Workers pick over sun-dried coffee berries near Goroka, New Guinea. Identify the coffee bushes and the shade trees, and note their relative heights. Note the material used under the drying beans, and the provision made for protection against thunderstorms.

b. Soils

Although coffee is fairly tolerant of a wide range of soils, it prefers a deep well drained loam with a high humus content and a slightly acid reaction (pH value of 6–6·5 is considered optimal). This depth of soil is most important due to the long tap root, while the surface-feeding fibrous roots readily attack raw humus in the topsoil.

4. Plantation Practice

Preliminary working of the land is very important. Although coffee will tolerate a wide variety of soils—provided sound agricultural methods and fertilisers are used—it naturally prefers a deep humus enriched soil. Initial clearing and burning of debris is followed by deep ploughing. The area is pegged out and the seedling trees and shade trees are planted.

The trees do not bear for six to eight years, and during this time the tree is pruned to about 12–15 feet, enabling picking to be carried out more easily. Without this pruning the trees would grow to around 30 feet. The coffee berries ripen unpredictably, so that trees may be picked three or four times in a season as the berries ripen bright red. Hand labour thus becomes very important in the picking stage and the development of a selective harvesting machine seems impracticable.

Yield per tree is low; usually around one pound might be the average, although there are wide variations in amount of green weight of berries picked. After picking the beans have to be fermented in tanks to remove the seeds from the pulp. They are then ready for drying. This process is normally carried out in the open, the beans being spread on hessian or canvas, and frequently raked over to ensure even and thorough drying. When they are thoroughly dry a considerable amount of hand sorting is carried out to remove undersized and damaged beans which would affect the final quality.

Finally the beans are hulled from the surrounding dry parchment and polished. These processes may actually be carried out on some plantations, but those that lack such facilities use the machines of a neighbouring plantation or send the produce to a central mill.

The coffee beans are then bagged for transport to the exporting outlet, and are at this stage ready for grinding to produce a finished product.

5. Operating Costs and Returns

Some typical averages over a number of holdings serve to point out the high capital costs involved in Plantation Agriculture. Total capital invested (all figures 1961, rounded to A$) on plantations ranged from $20,000 to $142,000 per unit. More meaningful were the figures per productive acre, which varied from $756 to $1,724 on land which was considered to be developed from a virgin state.

Annual yields as balanced against this varied from 360 to 2,540 pounds per acre, with an overall average of 1,150 pounds. Financially, the return on these averages ranged from $124 to $970 per acre with an average of $416.

A breakdown of costs indicates the high percentage involved in native labour charges, a significant factor in any Plantation activity:

Outdoor drying of coffee beans in the Eastern Highlands of New Guinea.

Average Gross Operating Costs per Acre—30 Holdings

Item	$	*Per cent*
Native labour	91·00	39·6
Other labour	16·00	7·0
Vehicles and Machinery	22·00	10·0
Maintenance of Structures	9·00	4·2
Hand tools	1·00	0·6
Fertilizer	27·00	12·0
Bags, Twine, Stencils	3·00	1·5
Insurance	1·00	0·5
Office expenses	4·00	1·8
Land rental	0·60	0·3
Freight	3·00	1·2
Undefined	11·00	9·3
Depreciation—on structures	4·00	3·8
—on plant and machinery	9·00	8·2
Total	$201·60	100·0

A budget for establishing and maintaining a holding could assume the following:

1. The holding would be producing coffee for the Australian market.
2. The holding was leasehold.
3. Two blocks of about fifty acres each would be planted out in the first three years.
4. The plantation would aim to complete a processing plant in five years, when commercial quantities would commence.
5. Provision would be made for access roads, living accommodation, store facilities and the coffee nursery.
6. Vehicles and hand tools must be purchased and maintained.
7. Year eight would be the first full production year, and the aim would be about 1,200 pounds per acre.
8. Expected price of processed and bagged coffee beans would be 35 cents per pound.

With such a budget of operating costs, and continuing export prices, most plantation operators cannot expect income to exceed costs until year seven or eight. The future of the industry continually looks to readily available capital to carry the plantation to this end, when a 10–15 per cent return on capital is necessary.

6. The Future of the Coffee Industry

Barring a radical change in the market outlook, the only long-term growth seems to be a slow expansion aimed towards the Australian market. The promising growth of the 1950's in this infant industry has slowed considerably. Coffee still remains the Territories' most important money-earner for the indigenous Highland population. The 1962–63 export income was over $2 million, accounting for about 11 per cent of total export income. Perhaps the most promising outlook with regard to development of coffee is the change from plantations to small holdings, where a mixed type of agriculture, including coffee, may prevail. In 1960 native growers operating small groves accounted for only about 25 per cent of total coffee output, although they held some 60 per cent of the total acreage. Estimates for 1970

Workers load coffee at Goroka airfield, destination Australia. Air transport is becoming increasingly important in New Guinea because of the difficult topography.

are that more than half of the proposed 12,000 ton production will come from native producers, and that small holdings with less than total reliance on coffee will be much more significant.

Bibliography

Aschmann, H., *A Consumer-Orientated Classification of the Products of Tropical Agriculture*, Economic Geography, 1952.

Aschmann, H., *Plantation Systems of the World*, Social Science Monographs, Pan-American Union, Washington, D.C., 1959.

Dwyer, R. E. P., 'Coffee Cultivation in Papua and New Guinea,' *Papua and New Guinea Agricultural Journal*, July 1954.

Dwyer, R. E. P., *The Coffee Industry of Papua and New Guinea*, Bureau of Agricultural Economics, Canberra 1961.

Fisk, E. K., *New Guinea on the Threshold*, A.N.U. Press, Canberra 1966.

Hessler, W. H., *'Coffee: Cash Crop of the Tropics,' Farm Quarterly* Vol. 18, 1963, p. 50 ff.

Howlett, D., *A Geography of Papua and New Guinea*, Nelson, Melbourne, 1967.

Laut, P., *Agricultural Geography:* Vol. I, *Systems, Subsistence and Plantation Agriculture*, Nelson, Melbourne 1968.

Questions

1. What are the main environmental requirements for the production of plantation coffee ?

2. List the cultural practices necessary to maintain adequate production levels on a coffee plantation.

3. Discuss the possible repercussions on Papua-New Guinea coffee if:

(a) A World Coffee Agreement forced Papua-New Guinea into world competitive trade.

(b) The present economic tie to Australia is maintained.

(c) Beverages such as tea suddenly replaced coffee drinking habits in Australia, or in the world.

4. How far is coffee growing in Papua-New Guinea typical of other plantation agricultural systems ?

5. 'It has been said that plantations are an anachronism and are proving a hindrance to economic and social advance in many emergent nations, yet New Guinea has only recently embarked upon this system of agriculture as a viable means of economic advance.' Critically discuss.

8 Pineapple Growing in Hawaii

1. Introduction

The pineapple is believed to be native to Brazil. Although it is not possible to find when pineapples first reached Hawaii, 1813 is the first year of recorded production. By 1886 the foundation of the modern industry had been laid, mainly through the effort of Captain John Kidwell. Kidwell, an English horticulturist had imported strains from Jamaica of the Smooth Cayenne variety, and this is still the principal variety used today.

Although the Smooth Cayenne had superior flavour to other types, it did not transport well, and it remained necessary for the establishment of the canning industry in the period 1899–1906 before the modern industry was firmly under way. Pioneers included T. D. Dale, A. W. Eames and W. P. Thomas, and by 1906 the modern industry was thriving.

An aerial view of part of the Dole Company pineapple plantation in Hawaii. The geometric pattern results from making use of all available areas of arable soils and the need to protect against erosion by planting on the contour. The central forested area gives an impression of the natural landscape. In the right centre, around the edges of some fields, planting material brought from the nurseries is ready for sowing.

The first commercial pack of 1,893 cases was produced in the year 1903, and this has grown in the 1960's to over 30 million cases of fruit in an average year. The industry employs over 7,500 permanent employees, although this rises to over 22,000 persons in the summer canning season. These employees produce an annual output valued at over $130 million. Total acreage is currently over 80,000 acres, and Hawaii accounts for over 75 per cent of pineapples entering international trade. This figure represents only a small amount of pineapples, since probably less than 10 per cent reach international trade—by far the bulk of the crop is grown for consumption in the United States.

In Hawaii the industry is located on five main islands—Oahu, Maui, Molokai, Lanai and Kauai. Here are situated six pineapple companies operating fourteen plantations and nine canneries. The size of the plantations is obvious when a comparison is made with total output figures.

2. Conditions Required for Pineapple Growing

The pineapple is an herbaceous perennial which is tolerant of lower rainfall conditions than those needed for many other plantation crops. It is propagated by suckers, crowns or slips set out in rows. The plant has long sharp leaves which spiral around a central stalk to a height of three to four feet. In eighteen to twenty months, depending on local conditions, the fruit is produced and ripens. After the single fruit is picked, two suckers at the base are retained and these yield more fruit in a further twelve months. Thus a ratoon crop is produced in the same style as with sugar cane.

A mechanical long arm harvester aids farm hands on Oahu Island, Hawaii. Note the race of the workers concerned and the number of rows being picked at one time. Note the topography of the pineapple fields in which this ratoon type crop is being grown.

Selected Statistics (*temperatures in degrees F.; precipitation in inches*)
Honolulu, Hawaiian Islands (21°N; 157°W.)

	J	F	M	A	M	J	J	A	S	O	N	D	Year
Temp.	71	71	71	73	75	77	78	78	78	77	74	72	75
Precip.	3·8	4·2	3·8	2·2	1·8	1·1	1·3	1·4	1·5	1·9	4·2	4·2	31·4

a. Climate conditions

Although rainfall conditions needed are only moderate—around 25–50″ per annum, an even distribution is required to maintain growth. Temperature conditions must be warm throughout the year and the modern mechanised pineapple industry increases and maintains soil temperatures by using black polythene sheeting. This black sheeting absorbs heat and at the same time helps maintain moisture in the root zone as well as discouraging weeds.

b. Soils

Apart from wet (poorly drained) soils, the pineapple is not greatly restricted by soil types. Hawaiian soils are generally highly fertile, decomposed volcanics which have weathered to a deep clay. They are heavily fertilised in modern commercial plantations to maintain this fertility, and soils are also fumigated to protect growing plants. Soil preparation is not very different from most other crops, it being

Hawaii's nine pineapple canneries each year pack over 30 million cases of pineapple fruit and juice. Machinery used in processing is modern and efficient as are machines used in all other stages of production. Here the freshly picked fruits are being fed into a machine which cuts off both ends, peels and cores the pineapples in one operation. The care in breeding plants of uniform shape and size, and with approximately the same ripening date, is of great importance in permitting widespread use of mechanisation.

ploughed, disked and brought to a fine tilth. In some areas simple terracing is carried out to preserve moisture or carry off excess.

3. Cultivation Practice

a. The Growing Period

After the slips are planted the fields require careful attention. The planting is still largely carried out by hand labour—plants being inserted in a hole poked in mulch paper, or in slits in the polythene sheeting. The amount of labour is considerable when it is noted that from 15,000 to 18,000 plants are inserted per acre. Newer mechanical planters are being developed, and at present are capable of planting more than 100 plants per minute.

Early attention includes surface fertilising, or alternatively foliar sprays. Considerable attention is devoted to protection from disease, insects and other pests by using highly developed boom spray trucks. Few agricultural activities have developed to such a degree of mechanisation.

b. Harvesting

Harvesting can only be carried out by hand, since the fruit must be inspected frequently so as to permit ripe pineapples to be selected for canning. The pickers are assisted by long-armed harvest machines which precede the workers through the field allowing ten rows to be covered at once. The conveyor arm carries the fruit 50 feet across the ten rows to an overhead storage conveyor, and the fruit is transferred to trucks. This method substantially reduces labour and effort, and allows the ten pickers and six crewmen to harvest over twenty tons of fruit per hour. Picked fruit must be handled gently and transported quickly to the canneries, so as to maintain flavour.

When harvesting is completed the one or two suckers for a ratoon crop are left on the plant and other planting material is selected. One or two ratoons are permitted before the field is ploughed.

4. Some other Special Features

Science has always played a major part in the Hawaiian pineapple industry. The PRI (Pineapple Research Institute) and individually conducted research stations between them spend over $1,000,000 annually. They continually strive to improve varieties and techniques, and to control pests and disease.

The industry suffered a precarious decline in the 1930's due to pineapple wilt carried by mealy bugs. Spray techniques have brought the disease under control as have later chemical compounds used in fumigating to control soil organisms. Scientific cross breeding of existing varieties has produced several superior types which will gradually replace the Smooth Cayenne variety. The developments of newer canning techniques have also added to the overall efficiency of pineapple production.

Boom spraying units move through well established pineapple fields on Molokai Island. The crop is being sprayed to guard against disease of the semi-mature fruits. The geometric pattern of planting in order to permit utilisation of machinery along the straight rows is clearly visible, and tractor-drawn spraying and harvesting equipment is constantly being improved in the attempt to reduce production costs.

5. The Future of the Industry

Although Hawaii has a pre-eminent position in producing pineapple, competition is increasing. Large shares are being taken of the world market by Malaysia, Australia, South Africa, Mexico and Formosa whose chief advantages are suitable soils close to expanding markets. Often cheap labour and lower freight rates together with government subsidies have encouraged pineapple plantations.

Despite increasing efficiency, there are special problems in Hawaii. High taxation rates, including a special 2 per cent 'processing tax', now operate. Diversification of agriculture has reduced some leasehold plantation lands originally under pineapples. Canned and frozen juices of other fruits are also an increasing threat to Hawaiian pineapple production.

Bibliography

Alexander, T. W., *Economic Geography*, Prentice-Hall, Englewood Cliffs, N.J., 1963.
Courtenay, P. P., *Plantation Agriculture*, G. Bell and Sons, London 1966.
Gourou, P., *The Tropical World*, Longmans, London 1958.
Pickles, T., *The Wealth of the World*, J. M. Dent and Sons, London 1964.
Zimmermann, E. W., *World Resources and Industries*, Harper, New York 1951.

Questions

1. What are the physical requirements for the growing of pineapples?
2. What advantages has Hawaii as a pineapple producing area?
3. Explain the unique characteristics of pineapple growing in Hawaii.
4. What are the problems which are likely to concern Hawaiian pineapple growers in the future?
5. Why is it that pineapples from Hawaii account for such a large percentage of the fruit entering international trade?
6. In most plantation farming areas, associated processing is only partial; yet in Hawaii the pineapple is totally processed by way of canning by modern techniques. In what ways does this reflect the contemporary form of plantation farming? Can the Hawaiian example be considered unique? Does it give a lead towards advancement for other plantation economies?

Part Four

Commercial Dairying

9 Commercial Dairying

1. Introduction

Although there are many areas of the world where milk from a variety of animals is consumed—from goats, sheep, horses, buffalo and reindeer, for example—commercial dairying on an appreciable scale is restricted to Western Europe and the USSR, North America and Australia-New Zealand. These are the world regions which have high annual milk production figures in relation to their populations, and high milk output per cow.

Some other countries which do not fall within this category are nevertheless worthy of note. India's huge population consumes very little locally produced milk. While her output is slightly less than that of France (19,718,000 tons against

THE WORLD
SHOWING AREAS OF
COMMERCIAL DAIRYING

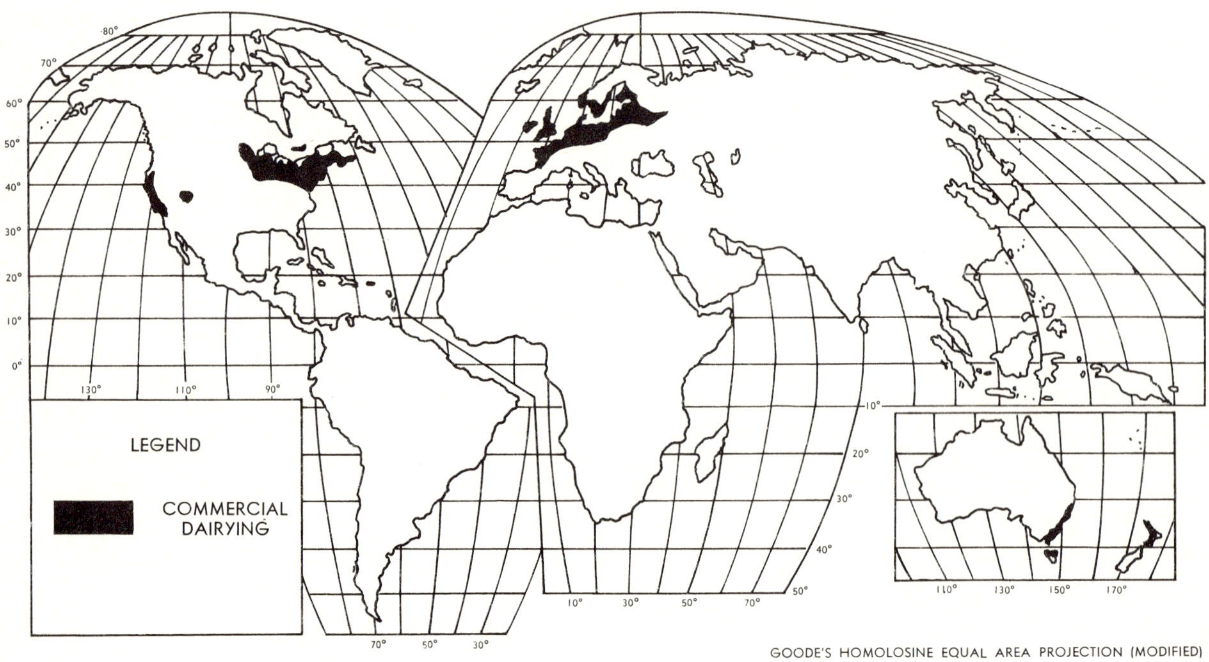

GOODE'S HOMOLOSINE EQUAL AREA PROJECTION (MODIFIED)

20,600,000 respectively), the yield per cow in India is the world's lowest, being only 480 pounds per annum. This has led to a great drive to increase production in India through the United Nations FAO Programme, and by assistance from foreign powers through schemes like the Colombo Plan.

On the other hand, while Israel's output is only 2 per cent of India's, the yield per cow is 9,640 pounds per annum—the world's highest above the Netherlands (9,500 pounds per cow). This is due to the highly efficient scientific methods the Israelis employ. Israel and the Netherlands are the only two countries where the yield exceeds 9,000 pounds per cow. The nearest others are listed below.

Table 1

Country	*Average Yield Per Cow (Pounds Weight) Per Annum*
Denmark	8,230
Belgium	8,160
Japan	8,010
West Germany	7,260
Switzerland	7,130
U.S.A.	6,430
Finland	6,320
U.K.	6,230
Sweden	6,230
New Zealand	6,120
East Germany	5,900
Norway	5,790
Canada	5,480
Austria	5,350
France	4,960
Hungary	4,910
Poland	4,490

(*Source:* Food and Agriculture Organisation Bulletin, 1963–64. Figures have been changed from metric measures.)

2. Basic Considerations

Although dairying is carried out in a wide range of environments with differing cultural backgrounds, there are several basic considerations.

i. Dairying is a Temperate Latitude Activity

Though mainly confined to Western Europe, North America and Australia-New Zealand with numerous diversified minor areas, the first point to consider here is the wide latitudinal spread of major world dairying belts.

a. North-Western Europe. Most countries of North Western Europe have considerable numbers of dairy cows. The following table indicates the ten leading countries, ranked in terms of output.

Table 2

Country	*Output Per Annum (Millions of Tons)*
France	20·6
West Germany	19·0
Poland	12·5
United Kingdom	11·5
Netherlands	6·5
East Germany	6·0
Denmark	5·5
Sweden	3·9
Belgium	3·8
Finland	3·3

Source: FAO Programme, 1963–64.

These countries generally have fairly high yields, ranging from a low of 3,430 pounds in Italy to 9,500 pounds in the Netherlands. Thus the latitudinal spread of these leading countries is within 60°N. Latitude—Finland and Sweden—to around 40°N. Latitude in Italy. Commonly the areas are coastal locations extending towards the interior. The major areas in terms of both total output and yield per cow lie between these extremes roughly in the northern section of 50–60°North.

Countries like the United Kingdom, France, West Germany and Denmark are all representative of typical locations. Here the farms occupy coastal and low-lying meadowlands of abundant precipitation, with warm summers and cool to cold winters. They are close to major markets and as well as being major producers of dairy products (butter, cheese and cream, and tinned, condensed and powdered milk) they are suppliers of whole milk to their huge urban populations.

Not only do the north-west European nations supply their own needs, but there is considerable movement of butter and cheese from country to country. Trade agreements, notably the European Economic Community (the 'Common Market'), have had a considerable effect in influencing quotas of these products in these temperate, highly productive dairy lands.

b. North America. In North America, as in Europe, there is a concentration around the large urban markets, in the Eastern States particularly. About two-thirds of the dairy cattle in the United States of America are found in a belt extending west from New York to Pennsylvania and the Great Lakes, into Michigan, Wisconsin and Eastern Minnesota. Portions of the Lake Peninsula and the St Lawrence lowlands of eastern Canada and areas of the Maritime Provinces are also important.

Thus most of these areas lie further south than those of Europe—40°N and 50°N. The orientation of districts towards urban centres is vital—thus New England, southern New York and Pennsylvania supply the eastern cities; south-east Wisconsin and northern Illinois supply Chicago and other major centres. Because of the large urban populations of the midwest, the interior boundary of the belt is found far inland along the southern shores of the Great Lakes.

Outside the eastern belt, California is the fourth state in total milk production (Los Angeles is the highest producing county in the nation) and there are other significant areas of production, particularly in the Pacific Northwest.

Obviously there will be a wide variation of environments in these areas. General characteristics are similar to those of Europe—abundant precipitation, cold to cool winters and warm summers. The exceptions are on the west coast—California with its hot dry summer raises dairy cattle through the feed stall rather than pasture.

c. Australia-New Zealand. Australia and New Zealand between them produce roughly the same amount of milk as the United Kingdom, and at roughly the same yield per beast. New Zealand's share of the total is about 45 per cent—a disproportionately high figure compared with her size and population, but a very effective yardstick to measure the importance of this industry, which makes up nearly 30 per cent of her exports. It may be difficult to substantiate the claim that New Zealand dairymen are the world's most efficient producers of milk, but she has one of the world's densest milk-cow populations and certainly ranks high among world countries in all aspects of the industry.

Since New Zealand's major dairy areas are in the North Island and Australia's are located on the east coast and in northern Tasmania, this region lies latitudinally equatorwards of North America from about 40°S to 30°S latitude. Here summers are a little warmer and winters milder than the other two world areas mentioned. Rainfall is abundant in New Zealand, though less reliable in her larger neighbour. In both countries mild climatic conditions make housing and shelter much less necessary than in the northern nations, and pastures are permanent because of this fact. Supplementary feeding is important, but is used to maintain all year yield rather than as an absolute necessity, as in Europe.

d. Dispersed minor world areas. It is virtually impossible to discuss other world areas along the lines of those above. India ranks second in world annual production, but has the lowest per cow figures—less than 6 per cent of the yield average in Israel per cow! Japan, on the other hand, has a high 8,000 pounds per cow yield, fifth in the world and a tribute to Japanese efficiency, though her total production is less than Mexico's inefficient industry.

Overall, wherever urban populations are dense, dairying is attracted, often overcoming environmental difficulties that stand firmly in the way.

ii. Physical Environment

A wide range of physical factors has produced a basic activity and diverse minor associations in dairying areas. The physical environment which suits dairy farming is also very satisfactory for a wide range of associated animal-plant activities. Included in these are fodder crops, primarily fed to the dairy herds and for farm storage or for sales when surplus permits; pigs and calves to consume portion of the skim milk residue; and vegetable growing.

a. Climate. Climate has already been discussed in broad terms. Some selected statistics are useful for further study and comparison:

Selected Statistics (temperatures in degrees F.; precipitation in inches)
Toronto, Canada (43° 39′N; 79° 20′W.)

	J	F	M	A	M	J	J	A	S	O	N	D	Year
Temp.	22	21	30	42	54	64	69	67	60	49	37	27	45
Precip.	2·8	2·4	2·4	2·3	2·8	2·7	2·7	2·8	2·7	2·6	2·6	2·5	31·4

New Haven, Connecticut (41° 18′N; 72° 57′W.)

Temp.	28	28	36	47	58	64	72	70	64	53	42	32	50
Precip.	3·8	4·0	4·0	3·5	3·9	3·2	4·3	4·5	3·7	3·7	3·6	3·7	45·9

Paris, France (48° 50′N; 2° 20′E.)

Temp.	37	39	43	49	56	62	65	64	58	50	43	38	50
Precip.	1·5	1·4	1·6	1·7	1·9	2·1	2·2	2·1	1·9	2·3	1·9	2·0	22·6

Boston, Massachusetts (42° 0′N; 71° 3′W.)

Temp.	27	28	35	45	57	67	71	69	63	52	41	32	49
Precip.	3·4	3·1	3·7	3·7	2·9	3·2	3·3	3·3	3·5	2·8	2·9	3·4	38·9

Auckland, New Zealand (37°S; 175°E)

Temp.	66	67	65	61	57	53	51	52	54	57	60	64	59
Precip.	3·2	3·9	3·2	4·4	4·8	5·4	5·5	4·4	3·8	4·1	3·3	2·9	48·9

In summary, then, dairy areas must have a fairly high yearly rainfall, usually over 30 inches, and high amounts in summer time which is the main growing season and also the time when the evaporation rate is the greatest. Temperatures must be moderate, so that pastures are reasonably uniform throughout the year and in particular with no poor pasture period. In most areas the pastures must be supplemented by stored fodder—hay and silage. A moderate, humid climate fosters forage grass and hay crops and allows a long growing season.

b. Topography and Soils. Topography and soils are both considerations in dairying, but because of the industry's reference to urban markets, frequently these two natural factors become less significant. As land costs rise in proximity to cities, dairying will occupy areas considered second rate when land was more readily available. Thus in the United Kingdom much heavy clay land formerly used for mixed farming is now used for dairying, although farms are usually located within easy reach of markets. The rapid, long distance transit of milk, often over distances up to a hundred miles, is characteristic. However, with the use of refrigeration, distances of hundreds of miles are not unusual. Places distant from the markets specialise in cheese, butter, canned and powdered milk, the best example being New Zealand-Australia for the United Kingdom market.

The considerations throughout dairying belts may be made in general terms. Topography does not have to be flat or gently undulating over all the farm unit, for pasture grasses do quite well on slopes and fodder crops can be grown on the better areas. Thus many dairy farms—in part at least—are quite rough compared with crop lands. The larger-sized farms illustrate the principle that whereas machinery cultivation can only be satisfactory on relatively flat land, slopes are quite suited to livestock.

In like fashion, soils need not be highly fertile. The better soils on dairy farms can be more profitably employed in crop production. Most pasture grasses will do quite well on poorer soils. Much of the world's dairy regions have been established on podsols, generally less fertile than the chernozems employed in cereal growing. Soils are often stony, which does not present serious problems, since little cultivation is carried out on slopes. The soils are frequently more poorly drained in dairy areas than they would be for crop production. The 'water meadows' in parts of Western Europe are extreme cases. Much of New Zealand's dairy belt has been extensively drained and yet retains fairly wet conditions unsuited to many other farming activities.

The basic requirement of dairy cattle is nutritious grass, the growth of which varies greatly from one area to another depending upon the climate. In New Zealand there is grass growth for practically the entire year, while in other areas such as Western Europe the cattle have to be kept indoors during the severe winters and are hand fed. Depending upon the growth and nutritive value of the grass sward, the diet of the dairy herds has to be supplemented with such additional food as beet-sugar mash and oil-cake. In areas such as Denmark the cattle consume little grass, being mainly fed on arable crops produced, with the aim of getting the maximum quantity of milk per cow and also the highest butterfat content.

c. *Associated activities.* While dairying is dominant throughout most of the belts outlined, there is a wide range of associated activities complexly inter-related. Some of these can be considered.

Fodder crops. Wherever cattle have to be wintered in barns, or in areas where pastures fall below certain productive levels, additional fodder must be provided for stall feeding. A wide range of vine and root crops are grown in many parts of European and American dairy belts, and less so in Australia-New Zealand, for reasons outlined.

Generally these crops are grown on the less steeply sloping areas of the farm, where soils are deeper and more fertile. Agricultural practices are modern and efficient. The crops grown are chopped by hammer mills, and other additions—hay, chaff, grain, molasses etc.—are frequently added. Turnips, pumpkins, mangel wurzels, sugar beets, swedes, and the green tops of these vegetables are also used.

Where warmer conditions exist, crops of maize, grain sorghum, oats, barley and other cereal types are grown. Apart from grazing in early stages of growth, the crops are harvested close to the ripe stage and treated in the same way outlined above. In some cases, such as in California, whole farms specialise in fodder production for cattle not grazed on pasture.

Pastures. Very few areas within the dairy belt have any substantial areas of native grasses. A very wide range of exotic grasses and legumes suited to particular climate-soil environments have been sown and are fertilised to maintain high quality, high persistency pastures. The addition of electric fences to control areas of grazing—'strip grazing'—assists in preventing pasture damage and soiling by the herd. There are no hard and fast rules for species of pasture plants, but some of the generally favoured species include clovers, rye grass, paspalum, lucerne (alfalfa), kikuyu, barley grass, bermuda grass, phalaris and many others in controlled combinations. Other considerations—danger of bloat, maintenance of pasture quality and quantity, and seasonal cover, together with fertilisers—create a complex pattern of pasture improvement.

Other animals. Pigs and poultry are often mentioned in dairy areas as other important animals. Poultry have little relationship to dairy farms in the modern sense of the industry, and pig raising has declined in many areas as greater emphasis is placed on grain feeding under highly scientific control. There is still an important place for skim milk in the feeding and fattening of porkers, and this sideline activity is a useful adjunct to milk production.

Many farms, especially those which buy small numbers of replacement heifers every few years, run a beef type bull with the herd, and the dairy cross calves are satisfactory veal propositions. This type of activity is generally restricted to marginal dairy areas or the more mixed type of farming where milk production is a smaller part of farm income.

iii. High Capitalisation

Size of farms and labour supply reflect an intensive highly capitalised industry with skilled management. Dairying is perhaps the most advanced type of farming. Milk cows, animal products and fodder crops need great attention if the quantity and quality are to be maintained. Scientific dairying gives a high yield, and with competition from such products as margarine increasing year by year all efforts must be made to produce the best quality products at the lowest cost. The time is rapidly approaching when only the most skilful dairy farmers will be able to compete.

Milking is the most regular job that has to be done on a dairy farm. The twice daily milking of cows means that they have to be kept on relatively small farms. The farms must also be within easy reach of urban markets if milk is to be sold daily.

Because of the great cost per acre of land and the need for expensive equipment, the capital outlay of the dairy farm is very high indeed. Much capital is needed for buildings to house cattle in very cold regions, and to store feed; and for buying implements to prepare the ground for crops and for harvesting, although this is sometimes done by outside contract these days. Milking machinery is also very expensive.

Although it is difficult to generalise on the size of farms, they usually lie between one and two hundred acres. On these relatively small units there are the three fundamentals outlined above—skilled management, high man–land ratio and high capital costs.

iv. Market Orientation

Few products are so highly market-oriented through such a complex transport system as dairy products. Because large quantities of whole milk are sold from dairies, the relationship between farm and market is extremely close. In many dairies close to large cities the whole production may be sold fresh, but beyond this there is a general decrease in whole milk through intermediate zones where butter and cheese become increasingly important.

This generalised pattern of concentric zones is a useful theoretical concept under which studies of typical representative farms can be made:

The inner ring is closest to the urban market and is the chief source of supply of whole milk. Usually a premium price is paid and some form of quota system may be employed. Because land costs are high in such proximity to the cities, the industry must operate efficiently and ensure a high quality product delivered fresh.

Outside this zone is an intermediate area where dairy farms supply limited quantities of whole milk to larger cities or for local consumption. The perishable nature of the whole milk will not allow it to be transported long distances without refrigeration. Distances, however, may be considerable. London, for example, draws milk from Somerset and Cheshire, both over 200 miles away, south-west and north-west respectively; while New York receives supplies from up to 400 miles distant. There is a strong tendency for butter and particularly cheese to replace whole milk as distance increases. The example from London can again be taken—close to London nearly 90 per cent of all milk produced is consumed as fresh milk, while in Somerset 75 per cent is manufactured into cheese and butter. Again, where populations are large enough to warrant it, milk will be moved in tanks on express train routes.

The third zone, theoretically an outer zone primarily concerned with butter production (cheese requires only 10 pounds of whole milk to produce one pound while butter requires 20 pounds to produce one pound), lies well outside major cities. Unless milk can be sold locally, the bulk of it is processed.

This zonation of dairy produce is well illustrated in the New York and Chicago metropolitan districts, and is repeated in many centres in Western Europe and in Australia. Where the population patterns are diverse the nucleation may become multiple rather than oriented towards a single market centre.

The degree of complexity of transport and correspondingly cost will lessen outwards from the zonal centre where less perishable cheese and butter is transported with less difficulty. Major export countries have developed intricate transport systems and handling facilities to ensure quality products overseas. This is particularly true of New Zealand, for example, which competes with continental dairy products on the world's largest consumer market, the United Kingdom. A word on the future of some aspects of transport offers interesting speculation as to long term effects on export-import relations in dairy products.

The dairy industry in the United Kingdom is due to inaugurate containerisation in its operations early in 1969. There may also be a co-ordination of similar services for Japan, the United States and some Asian ports. Australia will also follow this trend and New Zealand at a later date will initiate container facilities. This will involve changes in shipping as well as storage and handling facilities. Costs of insurance, feeder services, types of carriers and availability of labour will be involved.

It is expected that some benefits will result and if this means a lowering of costs for Australian and New Zealand dairy products it will be a major step forward. It is possible that a reduction of freight rates may result, as vessels designed for carrying containers should double the number of voyages per year and it has been estimated that five times the amount of cargo will be carried. Damage in transit should be cut to a minimum and this will be of great benefit to the dairy industry.

Costs should be cut or at least stabilised—a most significant factor for the dairy industries of Australia and New Zealand. This could be of great assistance in opening up the great potential market of East and South-East Asia. Containerisation in inter- and intra-state dairy movements may help in the competition with substitute products such as margarine. There may of course be labour problems and diffculties in the change to the mechanical handling of the containers, but containerisation is worth notice as it could be a most important development in a long line of important technical improvements which have gone into modern commercial dairying.

v. Production Trends and Problems

Within the dairying industry, these are a constant economic consideration. Some of the wide differences between production yields in various countries have already been outlined in this text and in tables, and the previous section indicates but one of the future possibilities. Others are outlined below:

Selection and breeding of the cows is of great importance and aimed at efficiency in the quality and quantity of milk produced. The principal dairy cows come mainly from the British Isles and Western Europe and include such famous names as Jersey, Guernsey, Ayrshire, Danish Red, Brown Swiss, Holstein and Friesian.

Others like the Australian Illawarra Shorthorn (AIS) are recent products of the 'new world'. Careful breeding and selection of cows has resulted in a steady increase in milk yields and butterfat content. It has been estimated that yields per cow have increased twenty times since the days of medieval peasant dairy farming. The use of artificial insemination has allowed the rapid development of first class dairy herds to the position where a cow can produce over 800 gallons of milk per annum. By using artificial insemination techniques, first class bulls can be used to inseminate enormous numbers of selected high producer cows—a figure of over 10,000 calves has been recorded from a single sire!

Without permitting deterioration of the products, modern developments of handling, storage, transportation and treatment have made operations much more economical. Mechanical milking is ubiquitous, as is the diurnal collection of milk in tankers in the most up to date dairying areas, together with the development of farm cool storage facilities and most recently proposals for the use of containerisation by such bodies as the Australian Dairy Board.

Many factors contribute to the economic risks that still remain in dairying. In the first instance, the excessive production in good seasons tends to work against the interests of the industry because local market saturation is soon evident. Yet the excess cannot be easily stored to compensate the farmer when an unfavourable season produces an insufficient quantity to provide a reasonable living. In economic terms the supply is inelastic. Resultant attempts to minimise the problem have seen the development of organisational structures such as the co-operatives of Denmark and New Zealand and the stepping up of tariff and trade agreements in the post-Second World War years as exemplified in GATT, of which both Denmark and New Zealand are members.

Bibliography

Alexander, J. W., *Economic Geography*, Prentice-Hall, Englewood Cliffs, N.J. 1963.

Boesch, H., *A Geography of World Economy*, Van Nostrand, Princeton 1964.

Jones and Darkenwald, *Economic Geography*, Macmillan, New York 1964.

Smith, Phillips and Smith, *Industrial and Commercial Geography*, Holt, Rinehart and Winston, New York, 4th Ed. 1965.

White, Griffin and McKnight, *World Economic Geography*, Methuen, London 1966.

Questions

1. Why is dairying considered to be one of the most intensive forms of commercial farming?
2. 'The location of dairy farms can be determined as much by economic as physical factors.' Discuss.
3. Explain fully the statement 'dairy products are highly market-orientated'. Give examples from each major dairying country.
4. Account for the great variance in yields per cow throughout the world.
5. In what ways is commercial dairying related to advanced urban-industrialised economies?
6. Carefully account for the typical zonation of dairying and any variance which may occur.
7. 'Although dairying is typical of highly commercialised farming, it is beset with many difficulties both on a local and world-wide scale.' Comment on this statement, using as many examples as you can to illustrate your points.

10 Commercial Dairying in Denmark

1. Introduction

i. The Background

Denmark is a small country (approximately 16,500 square miles) consisting of the Jutland Peninsula (9,000 square miles) and a number of islands such as Seeland (3,000 square miles), Bornholm (230 square miles) and Mon (85 square miles). The population is approximately five million.

Circumstances of the physical and cultural environment have led to Denmark's heavy dependence on the dairying industry for her economic survival. These include the introduction in the nineteenth century of cheap grain from North America and other countries including Australia and Argentina, which forced the Danes to concentrate on the production of meat rather than cereals. Then the introduction of refrigeration made the competition in meat from overseas too great also. As a result, the Danes decided to base their economy on dairying, the basis of which was intensive agriculture—arable farming rather than grass farming. Milk and dairy products became the main sources of income. Bacon pigs were developed later as an important subsidiary industry, which in a sense is a by-product of dairying. Specialised dairy farms averaging 50–60 acres, together with techniques of co-operative industry and indirect government subsidies, and development of industries directly or indirectly dependent on dairying, became the backbone of the national economy.

ii. The Present

The importance of dairying to the Danish economy can be seen when we note that Denmark produces one-third of the world's butter exports as well as great quantities of bacon, the production of which is directly dependent on the dairy industry for its skimmed milk.

Highly significant in the world pattern of dairying are Denmark's outstanding production figures, the world's highest for a major producer. These high-yielding farms are part of the agricultural organisation which produces about 50 per cent of all Danish exports; however, only 17 per cent of the workforce are agricultural workers.

The Danes have overcome the generally poor and infertile soils, and their efficiency has made the word 'Danish' or 'Danmark' highly respected in world markets.

2. The Place of Dairying in Denmark's Agriculture

No better example exists than Denmark of a nation which has achieved with great success a specialisation in agriculture which does not by nature suit it, but which has been made possible by the untiring efforts of her farmers and profitable by the

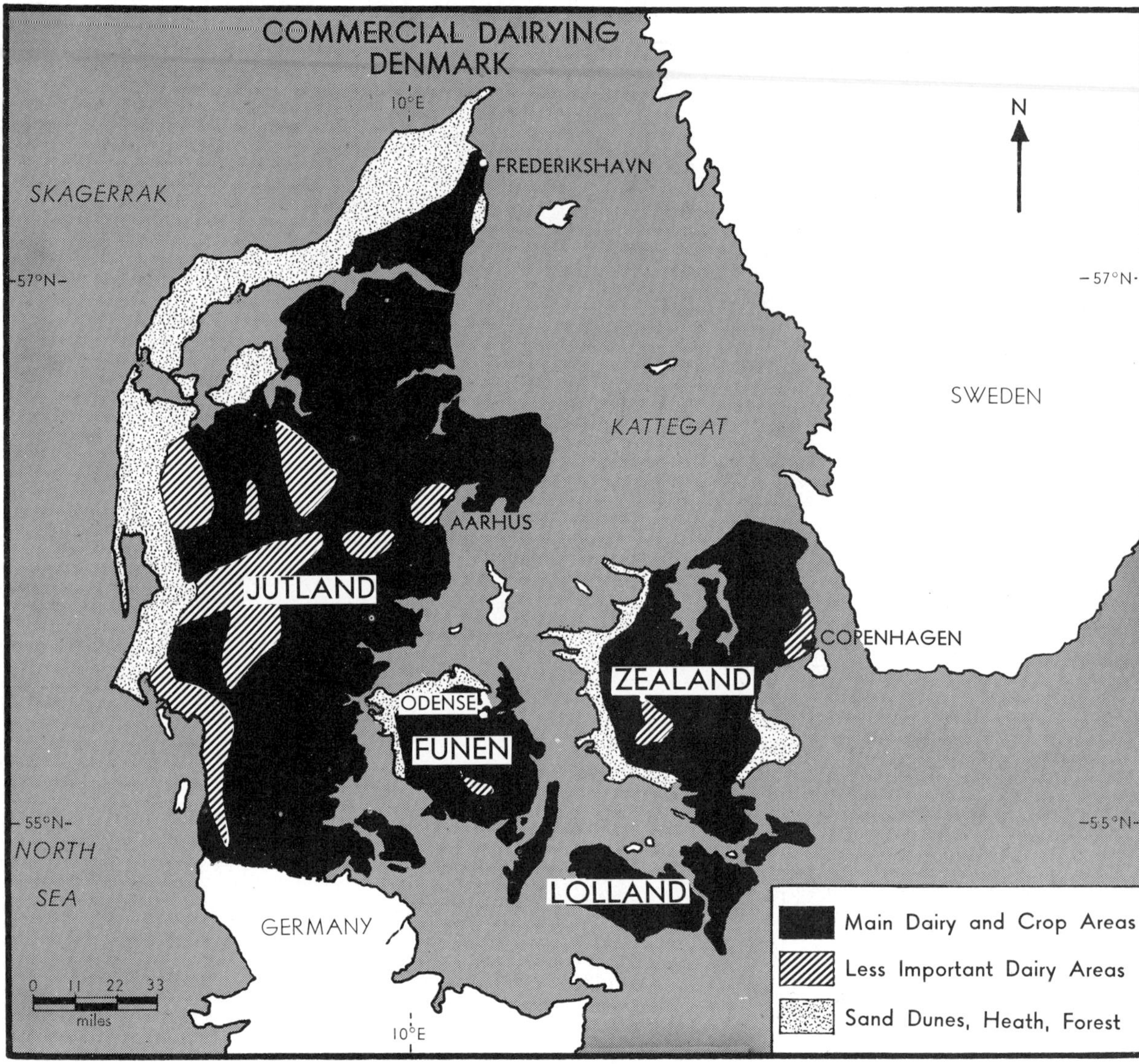

close proximity of the large urban markets of the United Kingdom and Western Germany.

It has been suggested that dairying in Denmark is to all intents and purposes arable farming: permanent pasture has been largely eliminated, and barley, fodder beets, mangolds, cabbage and kale planted instead. These crops are fed to the cattle, which are often tethered when in the open—this is especially the case in the south-east. This situation reflects the acute shortage of good land.

Denmark has been described as a large 'agricultural factory' which imports raw materials and exports the finished products. Some livestock feeds are imported and certainly they are some of the raw materials used in the dairy industry. In an industry which is dependent on the large urban populations of the United Kingdom and Germany for its existence, every effort is made to increase production and maintain the high quality of the products: a fall in quality could mean that these customers might look elsewhere for their requirements.

Labour-saving machinery is commonly used and considerable use is made of chemical fertilisers—ten to fifteen tons of commercial fertiliser per annum being used on the average 50–60 acre farm. In addition an average eight tons of liquid manure is used per acre to help grow root crops. Dairying in Denmark is highly organised and highly specialised; records are kept for every cow, indicating the costs of feed and the quantity and quality of milk yielded.

Magazines and government bulletins are studied closely to ascertain the latest methods and developments in farming and marketing. Scientific studies of problems over the years have helped build up the quality and quantity of farm produce. For example, since 1920 the crops per acre and the milk per cow have more than doubled.

Co-operatives have played a major role in the development of Danish primary and processing industries. They are voluntary organisations which buy seed, fodder and fertiliser in bulk and distribute them to their members. They collect, process and market the produce of the farms. The co-operatives do, in effect, cut out the middleman—the farmer virtually becomes his own middleman. The advantages are that the materials cost less, equipment needed on the farms is readily available when needed and there are no worries about selling products such as milk, bacon and eggs. The quality of the products is maintained by strict control and product grading. This ensures that the products have a top reputation. Variations of the Danish co-operatives are to be found all over the world—in Eire, Finland, New Zealand and California, for example.

3. The Natural Environment

i. Climate

Denmark has a maritime climate with prevailing Westerlies, and mostly experiences average temperatures around 40–50°F. and with an annual precipitation of 25–30 inches. The summers are warm and the winters cold, the coldest month being February which averages 30–40°F. As a result of the cold winters the cattle have to be kept in barns in winter and fed on sugar beet tops, cereals, roots and high protein oil cake. Of the 20–30 inches of rain which fall each year, two-thirds occurs between 1 June and 31 December—roughly summer and autumn. Thus the first half of the year is normally dry, unfortunately from the viewpoint of maize production which requires heavy amounts in the growing periods of April-June. The relatively high amounts in autumn also create problems, as maize is being harvested at this time. Frost-free periods are between 200 and 250 days per year, the longer period being experienced in coastal areas.

Selected Statistics (*temperatures in degrees F.; precipitation in inches*)
Copenhagen, Denmark (55° 41′N; 12° 35′E.)

	J	F	M	A	M	J	J	A	S	O	N	D	Year
Temp.	32	32	35	42	51	59	62	61	55	47	39	34	46
Precip.	1·5	1·3	1·4	1·4	1·5	2·0	2·4	2·6	2·1	2·2	1·9	1·7	22·0

ii. Topography, Soils and Drainage

Glaciation has played a major part in shaping the topography of the nation. The eastern section of the land contains vast areas of glacial deposition, and the soil type based on boulder clay is the most fertile. Most of the long coastline is fjorded and backed by low sandy areas. Even the central Baltic ridge seldom rises above 500 feet and most of the land is only around 100 feet. Many other areas are low lying peat areas. The islands, which are remnants of a partly submerged post-glacial period, are large in comparison with the 'mainland' of the Jutland Peninsula, and they too are low lying.

Topography has had a strong influence on land reclamation and in turn on soils. Less than a century ago one-fifth of the Jutland Peninsula was a waste of heath, bogs and marshes. Much of this has been reclaimed since the formation in 1866 of the Danish Heath Society (*Det Danske Hedeselskab*)—probably over two million acres of arable land up to the present time. Where topography and soils are unsuited to agriculture, afforestation has been practised; today Denmark has more than ten times the area of forest she had a century ago.

Not only lowlands have been drained and ditched, but inland and coastal sand dunes have been reclaimed. It is estimated that some 60,000 acres of drifting dunes have been arrested by windbreaks of trees. Tree planting (over 200 million trees in eighty years) has materially assisted in bringing shelter to exposed areas.

The Heath Society has also been instrumental in ditching, draining and dike building along the inlets and fjords to replace land lost by urban expansion and road building. The society co-operates with other agricultural organisations in soil testing and organising of machinery, fertiliser, etc.

This photograph shows the indoor use of barns. Note the prime condition of the cattle and the high standard building. This capital-intensive industry is efficiently run by experts with a high level of technological training. (*Photo courtesy of the Royal Danish Ministry for Foreign Affairs.*)

The picture shows Danish State small-holdings—small holdings set up by loan from the state towards the purchase of the land and the erection of buildings. The state small-holdings must have plots of a size to enable the farm to support a family. They are never less than eight acres and seldom more than thirty-eight acres. (*Photo courtesy of the Royal Danish Ministry for Foreign Affairs.*)

In summary then, Denmark's soils exhibit a wide range from heavy bog and peat types with generally high fertility, through fertile boulder clays and relatively infertile sands. In general terms there is not a great deal of 'easy' farming land—both in drainage and fertility. Continual rotation, extensive draining and heavy fertilising are constantly used to maintain and even increase the arable nature of the farmlands.

4. Farms, Herds and Production

In 1966 Denmark had over 177,000 holdings, the great bulk of them owned and occupied by individuals. Farm size is small to medium:

Farm Size	*No. of Holdings.*
Less than 25 acres	71,000
25–75acres	81,000
75–150 acres	21,000
Over 150 acres	4,000

This gives rise to an old Danish song that states 'few have too much land, and fewer still too little'. Since almost half the farms are under 25 acres, the holding is small enough to be worked by the farmer and his family without hired labour.

Tenant farms are few in number, accounting for only 5 per cent of all farms. The basis of farming is vested in individually owned smallholdings (less than 25 acres). However, figures indicate that small holdings are declining (from 80,000 in 1962 to 71,000 in 1966) due to amalgamation into larger units between 25 and 150 acres.

Herds are primarily of the Red Danish Cattle (RDM) or Black and White Jutland cattle, with Jerseys and Shorthorns being quite rare. Production figures are everywhere extremely high, as the following table indicates:

Breed	*National Numbers (1966)*	*Annual Yield per cow (pounds weight)*	
		Milk	*Butterfat*
RDM	356,980	9,867	421
Black & White	126,985	10,123	413
Jersey	76,200	7,590	458
Shorthorn	8,264	8,264	330

These figures are based on the year of 365 days, averaged over 46,250 herds totalling 561,000 cows in the 1966 report of the Danish Milk Recording Societies (*Kontrolforeningerne*). These figures are a significant sample of Danish cows from the nation's three million beasts.

If these figures are spectacular by world average standards, individual performances are more so. The 1966 annual record for production was made by a Red cow which gave 22,625 pounds of milk and 1,305 pounds of butterfat in a 365 day test!

5. Dairy and Allied Animal Products

i. Butter

Butter is by far the biggest source of income from Denmark's agriculture. Danish butter has played an important part in the international market since the first co-operative dairy in 1882—this followed the centrifugal separator invented in 1880 by L. C. Nielsen.

Tests on quality and food value are made each time milk is sent to the factory. These are based on hygiene and fat percentage. Cream is usually matured until the following day before churning into butter, to be packed immediately in different packs for export.

The Lur (Danish Bronze Age ceremonial trumpet) has been adopted as the symbol of the State Quality Control for butter, bacon, cheese and canned milk products. Only those dairies may use it whose products have been submitted to the most rigorous tests—kept fourteen days and then tested by nine experts. The right to use the Lurmark is taken away from any dairy whose product falls slightly below the standard required.

Much research has been done into the packaging of Lurpak products. Aluminium foil lined with parchment is used to ensure that the butter is protected from the effects of light and air and retains its churn-fresh flavour and quality. Danish butter is sold to one hundred different countries. It has such excellent keeping qualities under the right conditions that the Canadian Government kept one lot stored for two years! It was still in fine condition. It was marked as first class and it brought to the Danish dairying industry a compliment from the Canadian Government. Seventy-five per cent of Denmark's export butter goes to Great Britain—nearness of that country ensures ease of export.

No less than 75 per cent of the total area of Denmark is devoted to agriculture. Few countries can show an equally large percentage or such intensive utilisation. Large farms such as the one shown in this picture occupy less than one per cent of the agricultural area. Most of the farms are small or medium-sized. (*Photo courtesy the Royal Danish Ministry for Foreign Affairs.*)

Danish breeding cattle are in demand all over the world. Danish Red (pictured) have been bred since the 1840's. (*Photo courtesy the Royal Danish Ministry for Foreign Affairs.*)

Tests include taste, smell, consistency and also a chemical analysis. The judges' reports have a competitive significance as the results of each year's tests are published and the best and most skillful dairy managers get a valued prize plus an enhanced reputation.

Butter can be exported only by those authorised by the Ministry of Agriculture. Lurpak butter must conform to a number of strict regulations:

Only milk from TB free herds is used.
Butter is made of pasteurised cream, that is, cream heated to 176°F.
Butter must contain no more than 16 per cent water and at least 80 per cent butterfat.
No preservative can be used except common salt, butter is not to be coloured by analine dye.
Boxes and casks must show the date the butter was made and the identification number of the producing dairy.

Production of dairy products fell during the Second World War under German occupation, but production has slowly been built up since then and cheese exports have increased more than eightfold. In 1963, 80,000 tons of cheese were exported to many countries, with West Germany being the biggest customer.

ii. Cheese

The Danish dairy industry has expanded its cheese production to such an extent that it now stands third as an exporting nation after New Zealand and the Netherlands. Tradition played a large part in this expansion as the making of top quality cheese requires considerable 'know-how'. Danish cheeses are named after the particular district producing them and have a special character which has made them world famous.

Danish Blue is a great favourite in Great Britain and the United States. Some say the mould is green, not blue—the experts can debate this point, but the fact remains that this cheese is of splendid quality and is very highly regarded throughout the world. Samsoe, which was previously called Swiss cheese (no relation to Switzerland) is a truly Danish natural cheese which comes mainly from the island of Samsoe. This cheese is mild in flavour with characteristic regularly-distributed holes.

Only those cheeses with the Lurmark (a minimum 45 per cent fat content guaranteed) can be exported. The cheese from a particular dairy must pass rigid tests before the Lurmark can be used. Samples are taken at the dairies and at the cheese stores prior to export; any deficiency results in the Lurmark privilege being removed.

All hard and semi-hard cheeses now carry a special mark made from casein. It is then possible if a flaw is found to trace the cheese back to the dairy and the particular vat from whence it came. The mark is put on the cheese under the final layer of wax. It gives the percentage of fat, the word 'Danmark' and the production number. The final product is also marked with the stamp of the State Quality Control, which shows the control number of the dairy or exporter and the Lurmark.

iii. Milk

The Federation of Danish Dairy associations has made great progress in the production of high quality milk, largely through the initiative of the farmers themselves.

Danish cheese is produced under the strictest rules of cleanliness. The cheese dairies are very careful to maintain the high quality of their goods. This picture was taken at Marslev Cheese Dairy, Funen. (*Photo courtesy the Royal Danish Ministry for Foreign Affairs.*)

As well as the successful campaigns against TB and contagious abortion, the fight against mastitis is proceeding very well. Vaccination in threatened areas has helped also in the fight against foot-and-mouth disease.

Proper feeding techniques have been very important. First class milk results from the scientific mixing of winter fodder, the use of oil cake, silage, roots (especially beets) and hay. Production in summer months is only one-and-a-half times that of the poorest winter month. In recent years there has been a considerable increase in the condensed and powdered milk sold on the world market.

iv. Pigs

Pig raising is by far the next most important industry, and it is debatable whether pigs or dairying is Denmark's major export earner. The pigs are regarded as 'milk consumers', assisting to make the dairy industry pay, for they are fed large amounts of separated (skim) milk, butterfat and whey.

The Danish Landrace pig, a long-sided, light-shouldered but well hammed pig, is a tribute to careful and skilled genetic breeding. Progeny are tested and feeding techniques carefully formulated and analysed. The scientific techniques and testing employed in the dairy industry are carried out in pig raising.

Expansion has risen rapidly from the lowest level of 1,130,000 killed in 1942 to 6,700,000 in 1954 and 10,900,000 in 1964. The bulk of the trade is bacon, but there are extensive exports of canned meats, sausages, black puddings, etc.

One of the beautiful old farms of Funen, Denmark, with typical wooden framework and thatched roofs. The intense pride that is taken in both buildings and animals is evident from this photograph.

6. Pastures, Crops and the Farming Calendar

Crop rotation is practised everywhere in Denmark where conditions exist that are satisfactory to dairying. For example, in a farm of 80 acres, 65 acres may be divided into five fields of 13 acres each. No conventional fences are used; instead movable electric fences are used to control the feeding of the dairy cattle on different strips of cultivation. A typical five year rotation may be arranged as follows:

First year—improved pasture
Second year—improved pasture
Third year—oats
Fourth year—fodder roots (half swedes, half sugar beets)
Fifth year—barley and clover grass

The grass is used for the cattle to graze in summer, and as hay in the winter. The barley is used with skimmed milk to fatten pigs, the pig droppings in turn being used to fertilise the ground. The tops of the beets are used in silage for the feeding of the dairy cattle in winter.

There is a tendency which is increasing for the intense cultivation of the land to produce food such as roots and grain for cattle and pigs rather than pasture grasses. Fodder crops such as barley, oats, potatoes and other root crops take precedence over hay because they are more nutritious. Hay is more usual in areas which have lower fertility. A small area of a few acres may be divided into various strips each of which produces a different crop—barley, rye, turnips, wheat, oats and so on.

Outline of a Typical Dairy Farmer's Calendar
March-April—barley and/or oats sown, swedes and sugar beets sown, cows are brought from barns where they have wintered
May—crops are cultivated
June-July—hay making and storing
Mid-July, August—grain harvesting
September—ploughing and harrowing (for March sowing)
October—sugar beets dug, and put into field pits. The tops of the beets put into the silo pit
Early November—cows are brought in to winter barns
December, January, February—cows are 'hand fed' during this period by fodder, root crops, hay and straw.

7. Conclusions — The Future

Denmark exports two-thirds of its butter production to the United Kingdom, and cheese mainly to the United Kingdom and Western Germany. As a result it can be easily understood why the Danish government follows with great interest the endeavours of the United Kingdom to join the European Economic Community. Denmark has asked for special assurances from Great Britain that satisfactory arrangements be made so that the Danish dairy industry will not be too badly disrupted by Great Britain's entry into the Common market. The Danish government has also indicated that their country will also seek entry to the Common Market at a future date. Increasing tariffs on imports from countries outside the European Economic Community would strongly affect Denmark's export earnings.

Thus the problems which the dairying industry in Denmark now faces are primarily due to the changes which the EEC policy will introduce, and to social and political factors which have loosened family and national structure in Europe. The Common Market will further facilitate travel and intra-European migration and increase the possible drift from an industry whose character demands long working hours and close supervision. The possible entry of Britain into economic partnership in Europe could provide the problem of British Commonwealth product-dumping of excess dairy commodities. Obviously any upset to an industry which, as statistics presented indicate, forms the basis of national structure, presents the need for significant re-adjustment programmes. Increased automation is an alternative to labour shortages, but the initial costs would require heavy direct subsidy since the conditions of the EEC would prohibit protective tariffs. The possibility of a morning and evening milking supervision shift would certainly raise costs, although these might be offset by the attraction of labour to improved conditions and the promotion of local economic security through the multiplier effect of additional income-earnings.

The ingenuity displayed by the Danish community to meet the crises of the past suggest the problem is not insurmountable. At the same time, for the student of geography interested in the effects of the cultural changes indicated, the future of the dairy industry in Denmark presents an interesting socio-economic study.

Bibliography

Danish Agriculture, Department of Agriculture, Copenhagen 1966.
Larousse Encyclopaedia of Geography: Europe, Paul Hamlyn, London 1961.
Pickles, T., *The Wealth of the World*, J. M. Dent and Sons, London 1964.
Symons, L., *Agricultural Geography*, Bell and Sons, London 1967.
Wheeler, Kostbade and Thoman, *Regional Geography of the World*, Holt, Rinehart and Winston, New York 1964.

Questions

1. 'The dairy industry is more important to Denmark's economy than the sheep industry is to Australia.' Do you agree with this statement ? Qualify your view(s).
2. Discuss the locational factors pertinent to the Danish dairy industry.
3. In what ways does the Danish dairy industry differ from that of the other major dairying countries ?
4. What significance can be drawn from the fact that Danish dairy farms are essentially small but the yields per cow number among the highest in the world ?
5. What are the advantages and disadvantages of a dairy industry based on arable rather than grassland farming ?
6. Account for the rise, development and maintenance of the dairy industry of Denmark.

11 Commercial Dairying in New Zealand

1. Introduction

New Zealand's two main islands lie roughly between 34–47°S. latitude, giving them a more temperate climate than the major dairy lands of the Northern Hemisphere. Its area of 103,000 square miles is slightly larger than Britain. North and South Islands are each about 500 miles long and are separated by Cook Strait, only twenty miles at its narrowest point. The only other major island, Stewart Island, lies south of South Island, separated by Foveaux Strait.

Because of its historically later settlement, the land contrasts strongly with Denmark. The cultural pattern of land use reflects the rapid migrations of the eighteenth and nineteenth centuries and many parts are not far removed from frontier development. Isolation, especially from her world agricultural markets places New Zealand in a different position from Denmark. This, together with the environmental-physical differences, has created a distinctive pattern in the New Zealand dairying industry.

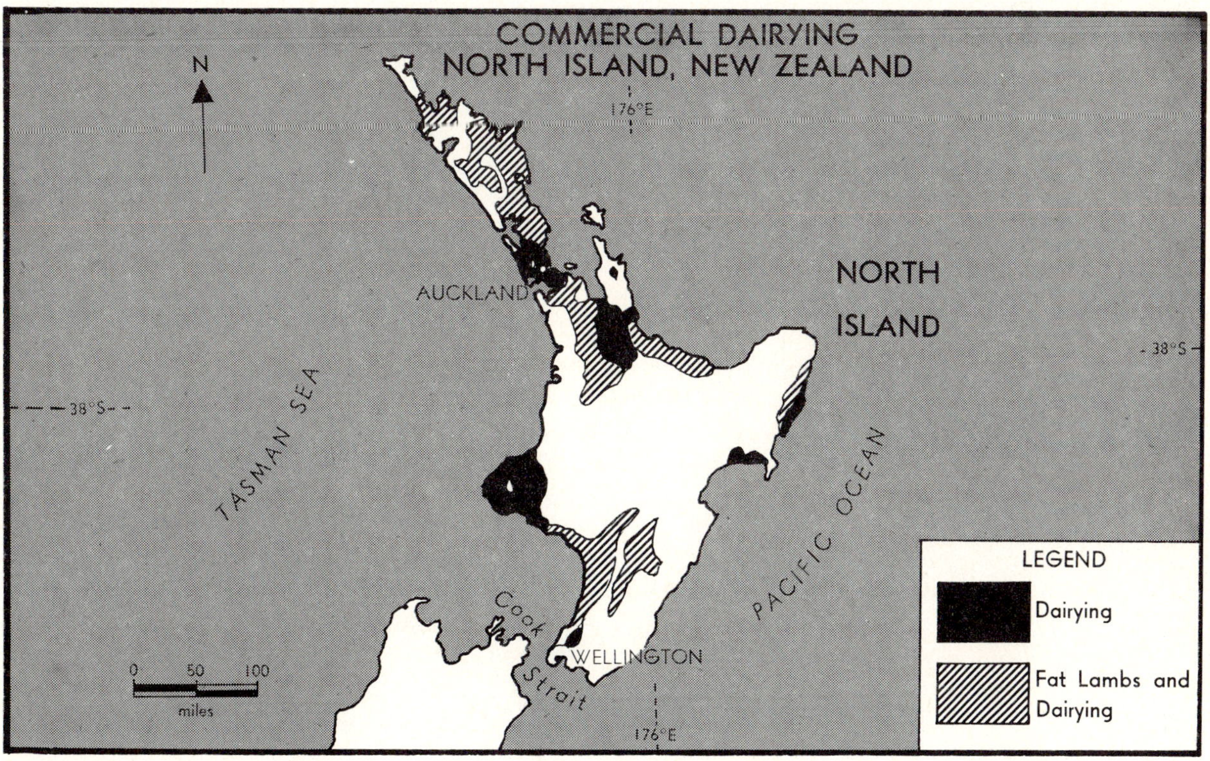

A high-class herd of Jerseys in typical Taranaki (New Zealand, North Island) country. This undulating country is ideal for dairying. Windbreaks and groves of shelter trees can be seen in the middle distance. The improved pasture contains a high percentage of leguminous clovers. In the right background field storage of meadow hay is indicated.

2. The Natural Environment

i. Topography and Soils

New Zealand contains elements of high mountains, rugged dissected hill lands and limited areas of plains. The dairy industry is restricted almost entirely to the North Island, which contains nearly 90 per cent of New Zealand's dairy cattle, chiefly located in flat lowlands or undulating lower hill lands. There are three major concentrations: the Taranaki–Mt Egmont–Manuwatu lowlands; North Auckland-Coromandel; and South Auckland-Waikato-Bay of Plenty districts. Here slopes are gentle and soils are generally fertile alluviums or weathered volcanics.

South Auckland-Waikato-Bay of Plenty has perhaps the greatest concentration of dairy herds in the world. This area produces approximately half of the country's total output. Taranaki has been a famous dairy area for many years. Northland and the Manawatu Plain are also important, although dairying is not so specialised in these areas. For example, the Manawatu area is more a fat lamb region than a specialised dairy farming region.

ii. Climate

Temperatures are warm to hot in summer and cool to mild in North Island's winter. Coupled with adequate and well distributed rainfall, ideal conditions are created for all year pasture. Temperatures around 40°F in winter are mild enough to mean that screen frosts (below 32°F) are rare—Hamilton recording only fifteen days of frost. Although rainfall has a winter concentration, sufficient of the 45–60 inches per annum fall in the dairy lands to allow summer pastures to flourish.

Selected Statistics (*temperatures in degrees F.; precipitation in inches*)

Christchurch, New Zealand (43° 30′S; 172° 35′E.)

	J	F	M	A	M	J	J	A	S	O	N	D	Year
Temp.	61	61	58	54	48	43	42	44	49	53	56	60	52
Precip.	2·2	1·8	1·9	1·9	2·9	2·6	2·6	2·1	2·1	2·0	1·8	2·5	26·3

Wellington, New Zealand (41° 16′S; 174° 48′E.)

Temp.	61	61	59	56	51	48	46	48	50	53	56	59	54
Precip.	3·0	3·3	3·2	3·8	4·9	4·7	5·4	4·7	3·8	4·2	3·4	3·5	47·8

New Plymouth, New Zealand (39° 05′S; 174° 04′E.)

Temp.	62	63	61	58	54	50	49	50	52	54	57	60	56
Precip.	4·5	4·0	3·7	4·8	6·1	6·2	5·4	5·8	4·9	5·6	4·7	4·6	60·2

3. Production and Management

i. Production

Whereas Denmark's dairy industry is based on arable farming with high reliance on stall feeding, New Zealand is quite the reverse—this is grassland farming almost totally.

New Zealand has a dairy cow population exceeding 2,000,000 with an average annual milk yield per cow of over 6,120 pounds. The national average size of each herd is seventy-eight cows and the country has about 35,000 dairy farmers. Approximately 32,000 of these farmers supply milk direct to the co-operative factories. The average annual butterfat yield per cow is nearly 300 pounds, making a total 550 million pounds of butterfat sent to the dairy factories. From this butterfat come the following products (figures for 1967):

Butter	240,000 tons
Cheese	106,000 tons
Milk Powder	115,000 tons
Casein	42,000 tons

New Zealand is the world's leading exporter of dairy produce and its yearly exports are made up approximately as follows:

Butter	195,000 tons
Cheese	96,000 tons
Casein	41,000 tons
Skim Milk Powder	75,000 tons
Buttermilk Powder	14,000 tons

Source: New Zealand Dairy Industry, NZIS., Wellington.

The dairy industry is one of the main sources of the country's overseas income. The industry makes up more than one-quarter of the total exports by value. Seventy-five per cent of the exports of dairy produce goes to Great Britain.

Cheese Making, New Zealand. After cheddaring the curd is milled. New Zealand Co-op Dairy factory at Waihi.

Fodder making from ryegrass pasture in the South Island. A feature of New Zealand's efficient pastoral and dairy industries is the attention paid to stored fodder for winter feeding of flocks and herds.

ii. Herd and Industry Management

Dairying has reached an advanced state of efficiency in New Zealand and such institutions as the Ruakura Agricultural Research Centre in Hamilton (the centre of the North Island's Waikato District) are world renowned. Overseas farmers visit such establishments to study the latest developments in management techniques and the methods and facilities that are used. Also, research work in the processing of milk into a wide range of products, and the development of new techniques with new machinery at the New Zealand Dairy Research Institute, have made the industry known throughout the world.

The entire dairy industry is organised around co-operatives. From the earliest days the farmers have pooled their resources to build factories to process their milk and cream. The co-operatives also market their own produce. The farmers, through the New Zealand Dairy Board, control the dairy industry from the cow shed to the overseas markets.

In breeds kept, Jerseys predominate, mainly because of the content of butterfat in their milk. They make up 80 per cent of the total, the rest being made up of Friesians, Ayrshires and Milking Shorthorns. Jerseys give extremely rich milk and under New Zealand conditions give a much higher yield of butterfat per acre than any other breed. By careful breeding to suit the New Zealand conditions, the quality of the Jersey cow has been developed to such an extent that the country has become the major centre for this breed in the world.

Recently, artificial insemination has become of great importance, two bull centres having been established in the Waikato and Manawatu areas by the Dairy Board. Semen is collected by specialists and sent to various areas throughout the country where technicians artificially inseminate cows on the farms. This is usually done during the early summer so that the calves are born in the following spring.

An elaborate system of herd testing is required and the bulls are selected on the basis of their daughters' production. A booklet on the New Zealand dairy industry published by the New Zealand Information Service, gives the following information:

Figures from herd testing units throughout the country are processed in the Dairy Board's Herd Improvement Department and, by reference to the electronically coded data, the performance of each cow under herd test can be produced at any time. These data are also used to produce statistical evidence of the quality of the bull. From those bulls which have produced female progeny with a yield consistently higher than the average for the herd, the best are selected for use in artificial breeding. As a result the few top bulls in the country can sire many thousands of calves each year, each one taking the place of about two hundred bulls needed under the old system.

Farmers keep a careful check on the amount and quality of milk produced by each cow so that the less able can be culled from the herd. The heifer calves from the best cows are kept so that the genetic quality of their herds is improved. Veterinarians have played an important part in keeping the herds clear of disease. A system of testing for tuberculosis has been introduced in all parts of the country. A shortage of veterinarians has led to the formation of veterinary clubs which farmers have set up on the co-operative principle.

The result of all this is that New Zealand, like Denmark, is world renowned for dairy products marked by the Grade Stamp of the New Zealand Dairy Division and Research Establishments.

4. The Place of Grassland Technology in New Zealand Dairying

New Zealand is representative of a developed nation whose progress can be associated with specialisation in a particular sphere, in this case grassland technology and the influence of Western economic growth. The latter has enabled New Zealand agriculture to gain access to highly urbanised and industrialised markets across the world. The extent of the New Zealand economy's specialisation in pastoral production can be gauged when it is realised that of a total arable area of 66·4 million acres, 68 per cent is occupied as farms, 28 per cent is sown pastures and 3 per cent as other improved farmland.

As an advanced economic system, New Zealand has shown that its agricultural operation is considerably less labour- and capital-intensive than similar operations in North America and Western Europe. At the same time the effort, expense and skill involved in the establishment of New Zealand pastures has been very great indeed. Lower labour and capital costs in New Zealand compensate for expensive freight charges to the market areas in the Northern Hemisphere. At the same time the materials were not by nature available. Improved pasture and grassland farming in New Zealand is the result of highly skilled techniques and extensive research. Nitrogen and phosphorous inadequacies have to be made up by establishing legume pastures after a superphosphate top-dressing.

Four stages have been necessary in the establishment and maintenance of pastures:

(a) Application of superphosphate (government subsidised);
(b) Sowing of specially developed red and white clovers;
(c) Further use of superphosphate and sowing of perennial pastures;
(d) Careful land and stock management including strip-grazing and control and balance of pastures.

Grasses saved during the autumn and pasture hay are used to supplement poor pasture growth in early spring. Grass silage is used especially in late summer and pasture hay is fed to animals in the winter.

5. Some Special Aspects of New Zealand Dairying

i. Locational Forces

In general, fresh milk for the urban market (for example, South Auckland supplies the city of Auckland), is produced in a zone quite close to the concentration of population. On the other hand, milk that has to go to dairy factories for processing into butter, cheese, ice cream, powdered milk and casein is produced in a zone which is more remote from the cities.

The market-oriented inner milk zone which is close to the cities and supplies the urban population makes up quite a small proportion of the New Zealand dairy products consumed. Milk, which has a low value per unit weight, goes to the market which is close at hand: the distance transported must be usually kept low.

The materials-oriented outer milk zone supplies butter, which has a high value per unit weight, to far flung markets, in New Zealand's case the United Kingdom and many foreign countries. In fact the United Kingdom takes more than 80 per cent of the butter that is produced in New Zealand.

The intermediate milk zone, which produces cheese, casein, powdered milk and condensed milk, involves a moderate weight cost reduction and is usually situated between the inner milk zone and the outer butter zone. In the case of New Zealand,

High quality Jersey milkers graze improved pasture at Pukekohe, Auckland Province. The cattle are rugged against cold despite the all year open grazing in this region. The well-ordered farm with neat buildings and rows of windbreaks typify New Zealand's highly efficient dairy industry. In the background are ewes and lambs grazed for fat lamb production.

a very large percentage of the products of this zone—for example, 90 per cent of the cheese and 98 per cent of the casein—is sent overseas also.

New Zealand processed dairy products do in fact occupy an outer materials-oriented zone in relation to the British market. The remarkable thing about New Zealand dairy products is that they can be put on the European market and can compete quite well even though they have to be transported across 12,000 miles of ocean. If import restrictions into the United States were lifted or relaxed, New Zealand dairy products would probably compete quite well there also.

ii. Effects of Administrative Policy

The Dairy Division of the New Zealand Dept of Agriculture keeps standards of products very high by the quality control system. Even the holds of ships are checked carefully to make sure that the goods are not likely to be damaged.

The Dairy Board decides what particular products to concentrate on in view of market trends—for example, casein, whey, butter, skim milk powder or sugar of milk, etc.

The decision to provide technical assistance in establishing processing facilities in foreign countries for recombining was an important one.

Dairy country near Taumarunui on the edge of river flats, and surrounding terrace and hill lands. More forest than is usual in much of the pastoral belt is obvious along the watercourses, but evidence of erosion in the background indicates a need to increase cover here.

iii. Recent Trends

When the New Zealand dairy industry was in its infancy the farmers pooled their resources to establish small co-operative dairy factories, which dotted the countryside. The invention and use of refrigeration has meant that dairy products can go much further afield and as the demand has grown so has the size of the dairy factories. This has been helped considerably by the use of milk tankers which have made it possible to collect milk from much wider areas. Whereas previously a small factory may have been supplied by a dozen or so suppliers, it is now usual for hundreds of farmers to send their milk to the one factory. The largest dairy factory has in the vicinity of 8,000 suppliers. However, all suppliers remain shareholders in their own dairy factory. Many small factories have gone out of existence, or have been amalgamated with others.

The great advantage of the large dairy factory is that highly-qualified professional staff can be employed and new techniques and equipment can be introduced more easily. The Department of Agriculture long ago implemented a system of grading for all export produce which assures buyers of the best quality.

Previously particular dairy companies specialised in one product or another—for example, butter or cheese. Now the larger ones have the facilities to produce a wide variety of products: butter, cheese, milk powder, condensed milk, casein, lactalbumin, lactic and anhydrous milk, all of which are exported. The New

Zealand dairy industry has expanded production in recent years, but the increased varieties of dairy products has made possible greater utilisation of the available milk.

iv. Trade Standards and Controls

New Zealand is by far the world's greatest exporter of butter—over 175,000 tons per annum—and the names 'Fernleaf' and 'Anchor' are brands which are known throughout the world. In addition, New Zealand is the world's second largest exporter of cheese. Great Britain is the major customer, although New Zealand exports to over sixty countries. Milk powders of numerous varieties and totalling 75,000 tons each year have particularly widespread markets. New Zealand supplies more than one-third of the casein that goes into international trade.

A most important development in the export trade is the recombining process: New Zealand butterfat and skim milk powder are imported by numerous countries, especially those in equatorial regions. Water is added to the powder to produce milk of the same food value and taste as normal milk. The Dairy Board's Technical Division and the Dairy Research Institute have done much research into this type

A milk tanker at a dairy farm in Kapiro. The photograph clearly depicts the closely subdivided nature of the farms with their well established hedges and pine wind breaks. The farmworker's cottage and milk shed are only part of a complex of buildings associated with dairyfarming. In the right middle distance is a protected silage pit, with the mixed herd of Friesan, Jersey and crossbred cows beyond. The bulk milk tanker makes a twice daily call at farms close to the Cambridge Co-op Dairy Company.

of processing and New Zealand dairy produce of this type has done very well on the export market. The New Zealand industry has supplied the best possible dairy products for recombining and also the technical know-how necessary to establish these enterprises in the importing countries. The technical services have helped greatly in establishing new markets.

The Dairy Board, which has thirteen members (eleven are elected by the directors of the dairy companies and two are nominated by the New Zealand Government) has the job of planning in accordance with the likely requirements of the market. The Board buys all the dairy produce for export, handles promotion and publicity, organises shipping to all markets and also looks after the complex job of accounting that is necessary for dealing with the $2 million worth of dairy products that are exported each year.

The New Zealand government does not subsidise the dairy industry directly, as is the case in most countries. Returns from the sale of dairy products have changed little in the last decade, and as a result increasing costs have had to be met by increased efficiency. It is a tribute to the efficiency of the New Zealand dairy industry that it is able to compete, and compete successfully, on the United Kingdom market with countries which are very much closer.

6. Two Sample Areas

The Taranaki-Manawatu Lowland and the South Auckland-Waikato-Bay of Plenty are two regions which exemplify much of the detail of the New Zealand dairy industry.

i. The Taranaki-Manawatu Lowland

New Plymouth's average January temperature is 63°F and its average July temperature is 49°F. The Westerlies bring heavy rain to this area: 60 inches per annum on the Taranaki lowlands and over 100 inches per annum on the hill country. The rainfall is well distributed throughout the year, a fact that is very important in the dairy industry. Further east on the Manawatu Plain the rainfall decreases to 40 inches per annum, and the farming emphasis changes from dairy cattle to fat lambs, although dairying is still important.

Upland Taranaki is shunned by farmers because of the difficulties of dissected landforms of deep valleys, high ridges and steep slopes. The lowlands, however, made up of material caused by the breakdown of the rocks (soft mudstones, siltstones and sandstone of the Tertiary age) are flat and with the addition of superphosphate are very good pasture lands.

Intensive farming is characteristic of the Taranaki lowlands with the small farm predominating. Each small farm has a high carrying capacity and is usually worked by a single family. Aerial methods of crop sowing, fertilising and top-dressing are much used to increase the fertility of the soil and the nutritional value of the pastures.

The farms encircling Mt Egmont vary from 50 to 140 acres in size. The rural settlement is dense, Jersey cows are the principal breed because of the high butterfat content of their milk, and small dairy factories dot the landscape. Further east between Patea and Wanganui fodder crops increase in importance. Hay and silage are, of course, important for supplementary winter feed. South of Mt Egmont small co-operative dairy factories specialising in cheese production are common. The tendency, however, is for the smaller factories to amalgamate or for bigger factories to be built in their place.

In the northern section of the Taranaki area between New Plymouth and Stratford, butter factories are more characteristic. These are bigger than the co-operatives to the south and draw their milk from a wider area. Cheese, especially Blue Vein Cheese from the Eltham area, is particularly important in the Taranaki area and is the main source of finance for dairy farmers here. Fat lambs are also raised, but this is a more important activity towards the east in the Manawatu Plain. When we consider that New Zealand exports 95 per cent of her production of cheese, the importance of this area in the country's export earnings can be better understood.

ii. South Auckland-Waikato-Bay of Plenty

This is a region of intensive pastoral farming with dairy cows relatively more important than in any other region of New Zealand. The climate of this region is ideal for dairying, with an annual rainfall of between 40 and 60 inches and plenty of sunshine. Thus pastures of rye grass and white clover grow practically all year round. The dairy farms are largely on reclaimed swamp lands and as a result peaty soils are common. Previously deficient soils have been made fertile by the addition of trace elements, especially cobalt. The pastures are largely man-made and are maintained at a high nutritional level by adequate and well-distributed rainfall, sunshine and very important top-dressing with superphosphates. The South Auckland-Waikato region produces more than 50 per cent of the country's butter and about 15 per cent of its cheese, also 50 per cent of its bacon and a considerable quantity of lamb. Ninety percent of the cultivation is for grass and the region produces 30 per cent of the country's hay and silage.

Mr G. W. Johnson's very modern milking shed, Waikato.

Mechanisation has been a very important factor in the development of this region—the use of machines for top-dressing, seeding and mowing is widespread. The use of light planes for aerial top-dressing is also a feature particularly in the more hilly areas. Systematic use of permanent pasture of high productivity has been made possible by the use of electric fences.

This is New Zealand's leading dairy area. The land is generally poorer than Taranaki, but high production has been achieved by the application of the latest techniques in drainage of swamp lands, and pasture development with the use of fertiliser and the correct seed selection. South Auckland-Waikato has long displaced Taranaki as the leading dairy region in New Zealand.

In the valley of the Waikato, New Zealand dairying is seen at its very best. Hamilton, Te Awamatu and Otorohanga are focal points for the dairy industry and the Jersey cow is supreme. It is, however, an area which is still not developed to its fullest extent, a fact that is no doubt determined by world markets. Further development may eventuate with the rising standards of living in South-East Asia.

7. The Future of the New Zealand Dairy Industry

The development of New Zealand agriculture and the Dairying industry in particular will depend as much on external trade factors as on the suitability of the physical environment. Britain's continued application to join the European Economic Community threatens the loss of a considerable market. Sixty percent of New Zealand's export trade went to the sterling area in 1966. On the other hand the Kennedy Round Agreement in 1967 between fifty non-Communist countries proposed an average cut of tariffs varying from 35 per cent to 50 per cent. A considerable boost to trade prospects between developed countries could result and balance the drop in exports and resultant revenue loss from Common Market developments. A further complication arising from the EEC could be a reduction in British capital investment in New Zealand. Loans involve interest payments and balance of payments must be adjusted in the long run by balance of trade. This would be impossible if the conditions of Britain's entry into the EEC involving non-preference for Commonwealth nations, were insisted upon.

Throughout its history New Zealand has been dependent upon overseas trade for its development and progress. The United Kingdom has continued to remain New Zealand's most important trade partner—buying 60 per cent of New Zealand's exports in 1966 and supplying 62 per cent of her imports. The sterling area takes almost all of New Zealand's cheese and butter. Dairy produce made up 25 per cent of total merchandise exports in 1966. Consequently the need to find new markets and develop new products has been a major preoccupation of New Zealand government policy and has introduced considerable marketing influences for New Zealand agriculture. A Trade Promotion Council established in 1962 undertakes trade surveys and missions and assistance to exporters. An Export Development Conference in 1963 led to an Export Guarantee Insurance Scheme. A New Zealand-Australia Free-Trade Agreement of 1965 proposed the gradual removal of tariff barriers or tariff reduction. The precautions taken by the Customs and Dairy Board to ensure high standards reinforce the significance of the export trade earnings which will help maintain and increase living standards in New Zealand by paying for imports and interest on capital loans.

The importance of export earnings derived from agricultural products was further illustrated at the 1964 Agricultural Development Conference, which initiated a programme for rural production to be reached by 1972. The targets were designed to ensure sufficient foreign exchange needs to enable a 4 per cent increase in real Gross National Product without the need for government borrowing abroad. The growth of the first three years was to be double the current average with slow growth in the next two years and high rate of growth in the last five years. The increase in post war livestock productivity through technical changes outlined was expected to continue. In dairying, butterfat production per milking cow was calculated to rise by one per cent each year. Results recorded between 1962–66 exceeded expectations—a 14 per cent increase in farm livestock included a 6·9 per cent increase in dairy cattle.

Features of world economic change as evidenced by GATT, EEC and more recently the Kennedy Round may well lead to further development of New Zealand as an advanced commercial society. Specialisation of production in the dairying industry is markedly evident where reliance is placed on Jersey breeds (80 per cent of the total) and high butterfat yields instead of diversification of production among sidelines such as pig-raising. The high incidence of co-operative factories, which involve 70 per cent of dairy farmers, has enabled the widespread application of technical developments and research staffs and has facilitated quality inspection and marketing processes.

Bibliography

Cumberland and Fox, *New Zealand—A Regional View*, 2nd ed., Whitcombe and Tombs Ltd. Sydney 1965.

Cumberland, K. B., *Southwest Pacific*, Whitcombe and Tombs, Sydney 1960.

New Zealand Dairy Industry, New Zealand Information Service, Wellington, 1967.

New Zealand Official Yearbook, 1965, Government Printer, Wellington.

Robinson, K. W., *Australia, New Zealand and the Southwest Pacific*, University of London Press, 1960.

Questions

1. Compare and contrast the organisation of dairying in Denmark with that in New Zealand.
2. How has New Zealand's dairy industry been able to compete on the European market despite her isolation ?
3. Discuss the advance in dairy technology as exemplified by the New Zealand industry.
4. Discuss the problems confronting the New Zealand dairy industry.
5. What physical and economic factors underly the systems of arable dairy farming as against those of grassland dairying ?
6. In what ways does the New Zealand dairy industry bring out the factors of location in commercial agriculture ?

Part Five

Crop and Livestock Farming

12 Crop and Livestock Farming

1. Introduction

This form of agriculture is based on both crops and animals. Crops are either produced for direct sale, or are grown to be fed to animals, the animals in turn being sold. Not only crops, but pasture too plays an important part in the activity, though on some intensive farms grass may be almost wholly replaced by a fodder crop or a combination of fodder crops. Since this agricultural activity lies between livestock ranching on the one hand and crop farming on the other, it has characteristics of both. The relative importance that the animals or crops play in the combination is also to be considered. It is possible to find all degrees of transition in crop-livestock farming.

Crop and livestock farming is a development from Western European agriculture, certainly if it is in its most sophisticated form. It is becoming increasingly important in the developing economies, as higher living standards allow the peoples to substitute animal protein for plant protein in their diet. Some form of animal-plant association may evolve in the underdeveloped economies in the simplest form, however. Thus throughout much of south-east Asia the pig is responsible for cleaning up spent grain or inedible portions and tops of field crops. As the animals root up the old plants and add fertiliser to the soil by way of their droppings, they can be considered part of the agricultural pattern, even though it may not be a deliberately planned attempt. The studies which follow are concerned with mixed farming which has been carried to a very high level of development. Here a deliberate attempt at planning ensures that crops and animals are in a direct relationship, where one works for the benefit of the other. There may also be a strong correlation between the economics of price and changes in consumer demand, which influences the animal to plant ratio on the farm and eventually in the market situation.

2. Location

Crop and livestock farming of some kind is carried out in all continents. Though there are as indicated above considerable differences in degree of intensity of plant to animal and differences in mode of agricultural activity, certain belts can be identified.

i. Europe

Here the belt extends far inland from the west coast with a wide latitudinal spread between approximately 40–60°N. The area is by no means continuous, being interrupted by croplands, dairying and other activities. The difficulties of classification in some areas are obvious. There may be overlapping with market garden and orchard areas where the influence of urbanisation is strong; poultry

THE WORLD
SHOWING AREAS OF
CROP AND LIVESTOCK FARMING

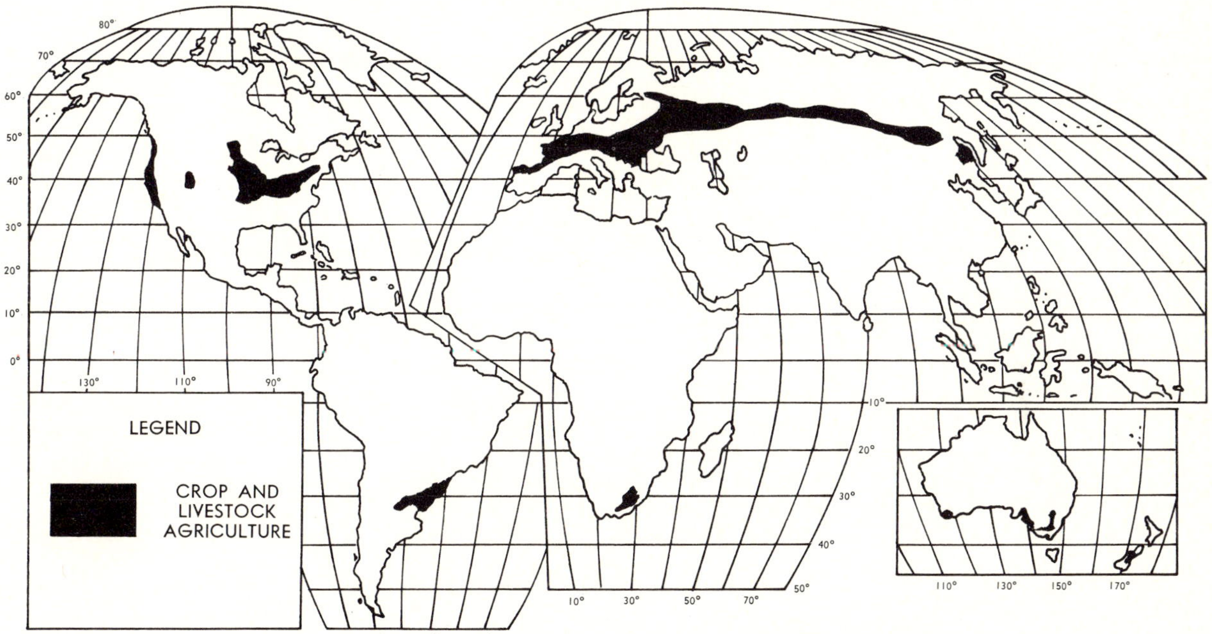

GOODE'S HOMOLOSINE EQUAL AREA PROJECTION (MODIFIED)

farming for both meat and eggs may also be important. Thus classification is easiest in small continuous areas.

ii. North America

Crop and livestock farming in North America covers extensive areas of the eastern and mid-west states. There is a belt also in the Pacific north-west states and smaller areas in other parts of the country. Again, the boundaries of the areas are difficult to delimit due to considerable overlapping with other agricultural types, and mapping on small scales is a problem. The latitudinal spread in North America is not as great as in Europe-USSR, but the belt lies about 35–45°N. in the eastern states. Thus the climatic conditions are rather warmer in the American areas, and there are considerable differences in crops grown and livestock raised. Basically, however, the animal-plant relationship holds, with the difference that the American farmers probably rely more heavily on the animal as the final product, whereas their European counterparts gain income from sales of both animals and crops. Processing by canning, drying and freezing is also more widespread in North America, due mainly to increased distances to market.

iii. Other World Areas

The distribution of other world areas is quite scattered. Generally they lie in the middle latitudes around 30° to 45° in the Southern Hemisphere. In South America, the coastal areas of southern Brazil and Argentina and the south of Chile are the main regions. In Africa there is an extensive area of the central south extending practically from the tropic to the coast. The other major area is in Australia-New Zealand. In Australia the temperate interior of the south east is the major belt, while there are small scattered areas in both South and North Islands of New Zealand.

3. The Physical Background

i. Climate

This agricultural type is found under a wide range of climatic conditions. The length of the growing season may vary from a short 180 frost free days in parts of the USSR, through 240 frost free days in much of the American midwest, to over 300 days in the more favoured parts of Argentina. These variable conditions permit the growing of a wide range of cereals, and most livestock types are raised in conjunction. Conditions are difficult to discuss generally, as are the variations in land use which result. Coastal margins lack the temperature extremes of interior locations, and here maize may replace the small grains of colder areas—rye and barley, for example—while wheat and oats are found almost everywhere. These conditions also affect farm size and in turn size of farms greatly influences livestock types. Smaller intensive farms favour porkers, or vealers and yearling beef. Where wheat is dominant, fat lamb raising and wool growing are the major associated activities.

Winter temperature conditions may range from as many as six months below freezing to no months with an average below freezing. Summer conditions may also show a wide range. In the cereal areas summers are long, cloudless and hot. Monthly temperature averages thus may vary from a low 30–40°F. to over 70°F., winter to summer. The selected statistics for typical stations give some idea of the ranges experienced.

Precipitation limits the crop types grown. The lower limit for effective small grain cropping is about 15 inches per annum, and the upper limit is around 50 inches per annum. These figures correspond approximately with the lower limit for wheat (15 inches per annum), while 50 inches per annum sets the boundary between maize and wheat. Distribution of precipitation is important; winter-spring rainfall and a summer dry ripening period is favoured for cereal cropping.

Selected Statistics (*temperatures in degrees F.; precipitation in inches*)

Columbus, Ohio (39° 57′N; 83° 02′W.)

	J	F	M	A	M	J	J	A	S	O	N	D	Year
Temp.	29	30	40	51	62	71	75	73	67	55	42	32	52
Precip.	2·9	2·1	3·4	2·7	3·1	3·3	3·5	3·2	2·6	2·4	2·2	2·6	34·0

Berlin, Germany (52° 32′N; 13° 25′E.)

Temp.	31	32	37	46	55	62	65	63	57	48	38	33	47
Precip.	1·5	1·3	1·7	1·4	2·0	2·0	3·1	2·2	1·8	1·8	1·6	1·7	22·1

Warsaw, Poland (52° 12′N; 21° 0′E.)

Temp.	26	29	35	46	57	63	66	64	56	46	36	30	46
Precip.	1·2	1·1	1·3	1·5	1·9	2·6	3·0	2·9	1·9	1·6	1·5	1·5	22·0

Sao Paulo, Brazil (23° 38′S; 46° 37′W.)

Temp.	69	69	68	65	60	59	58	59	62	63	66	68	64
Precip.	9·1	8·5	6·2	3·0	2·9	2·2	1·3	2·1	3·4	4·5	5·2	7·6	56·0

Another major consideration is the danger of unseasonable weather. This danger is more apparent towards the interior periphery. In the warmer world regions, careful conservation of fodder, pasture and water resources is very necessary. Prolonged dry spells or over-wet periods, late summers and heat wave conditions may all present problems at times. The crop and livestock zone can alleviate much potential danger by careful husbanding of resources, and these areas of mixed activity are at least 'safer' than the cereal cropping or livestock ranching belts of the continental interior. In the cooler temperate and cold temperate belts, the dangers may take the form of unseasonable, late or severe frosts. The ripening and harvesting of crops may be delayed. Under such conditions, much capital must be directed towards shelter sheds, barns, fodder storage facilities, grain drying devices and so on.

ii. Topography and Soils

Topography and soils throughout world regions are described under a variety of headings. In the United States, for example, the terms 'prairie-chernozem' indicate plainlands based on fertile chocolate–black earths. In Europe and the British Isles soils lack a single terminology, though they are commonly podsol types of varying fertility. 'Plains' or 'Downs' are acceptable topographical descriptions in these lands, also in Australia and New Zealand. In South America the humid pampas approach the American midwest in soil type, both being formed under a tall grass cover, and landforms are also flat to undulating generally. In South Africa the local term 'veldt' may be used.

The moderate to flat topography permits a wide range of machinery to be used throughout these crop and livestock lands. A high degree of mechanisation for all purposes—ploughing, cultivating and harvesting—is thus permitted. Although landforms do not influence the use of machinery to any degree in the animal raising aspects of these farms, certainly costs for shedding and storage facilities, for transport and communication routes (by rail, road or canal) are less costly than in more rugged areas.

Of much greater significance than the soil character is the degree of care exhibited by farmers in maintaining and improving textures and fertility through a wide range of activities. Included here is the addition of organic and non-organic manures, and the use of leguminous pastures to maintain fertility: dolomite liming to improve texture and a range of mechanical devices to assist soil aeration and improve soil tilth. To be considered here also is the involved pattern of rotation of crops and planned depasturing of stock—characteristic of this advanced form of agriculture. Ditching and draining to sweeten heavy soils, and green manuring to add organic matter to lighter sandy types is common where conditions demand such steps. In summary, a wide range of mechanical, chemical and organic measures in controlled combination allow intensive use of these lands, and a continuing or even increased output of vegetable and animal products.

As urbanisation makes increasing inroads into the mixed farming lands, size of farms decreases. Higher costs of labour, materials and administration necessitate improved farming methods to lift output per acre. In this form of agriculture, the maintainance of soil fertility and the increasing inclusion of second rate slopes and poorly drained flats into the farm units is characteristic. In few forms of agriculture is sound farming practice more demanding.

4. Basic Considerations

Although it has been indicated that there are difficulties of fine definition in crop-livestock agriculture, there are certain basic considerations which generally hold for all world areas.

i. Crop and Livestock Agriculture Reveals Stages in Diversification

The development of crop and livestock farming has been a gradual historic process, and it is only in recent decades that there has been a deliberate planned attempt to integrate animals and plants within a single agricultural unit. Certainly crops and animals were raised on the same farm in medieval Europe and this is still the case in many parts of the world, even where agricultural levels of development are considered low.

Modern crop and livestock agriculture, however, is a conscious aim at diversification of activities on either a supplementary or complementary basis. In the former case the activities together earn farm income and in the latter the end product—animals—are fattened on crops grown for this purpose.

The pattern of development of this form of agriculture has been observed in recent times in the plains of the 'New World'. The first phase was one of pioneering. The pioneers cleared large tracts of land in the virgin forests and grasslands and generally adopted a monocultural approach. The widespread wheat farming on the eastern prairies of central North America and the pastoral activities further west are typical. As yields fell due to the depletion of soils, cycles of dry years and the economic problems which ensued, the pioneers were forced into newer marginal lands. The continued occupation of virgin lands meant little real long term permanency.

A temporary phase ensued in which semi-permanent holdings were occupied. These large farms aimed primarily at a monocultural system, but there was considerable overlapping of large and small stock with grains—not so much on the same farm unit as side by side. Again, many of the units suffered from the exploitation inherent in this system of farm management, and some settlers deserted farms for newer pioneer lands.

The final phase of development saw considerable subdivision of holdings and a rational approach to conservation of the land resource. The necessity to provide a grass cover in fallowing wheat lands (a natural corollary of rotation methods) and the introduction of grasses in the grazing lands where natural species failed to regenerate, brought about a pasture revolution. Thus a pattern of permanency was forced into the land use activities. With stabilisation on these smaller units, diversification was natural. Cattle and sheep were introduced to cope with the flush of pastures. Small grains shared in a rotation pattern with improved pastures. The further diversification of activity is a comparatively modern development.

Such diversification is a result of several factors:

a. Economic considerations. As prices fluctuate for cereals, meat and wool, there is an inverse correlation created. Wheat acreages expand as beef, mutton and/or wool prices fall. As market demand weakens due to the increased output of grains, there is a tendency for livestock numbers to increase. The economic relationship is not completely balanced, for although there is a time-lag in building up livestock numbers, cereal production can be expanded in one year. It requires a number of years before livestock breeding can raise numbers, and during this period stock prices may rise with increased demand, and there may be considerable movements of store stock from other areas. These areas are the zones where breeding is the major activity, rather than fattening—that is, the livestock ranching zones which are generally climatically less favourable. There is, however, a clear lesson, that of combining stock and crops. The old adage of not carrying all the eggs in one basket may hold true.

b. Climatic considerations. The previous section introduces another factor—that of climatic variation. Diversification is an obvious method of counteracting, in part at least, the vagaries of climate. Although both stock and crops will suffer in years of below average rainfall, the risk of total failure can be minimised. If, for example, autumn and spring rains are inadequate to produce satisfactory early growth in wheat or oat crops, stock may be turned in to eat off the young growth. Generally the stock can be maintained in store condition throughout prolonged dry spells, providing water supply is adequate, and provision for stored fodder has been satisfactory.

Crops of course will fail in varying degrees during periods of drought, but quite satisfactory yields can be obtained when rainfall is well below average, providing the distribution is at the right times. After the drought breaks the crops may become the major source of income in the first season, especially if livestock have been heavily depleted, and restocking is too costly. In these instances the agriculturalist may make nearly all his income from grain—the time-lag previously mentioned being against animals in the first one or two seasons at least.

c. Planning. A later chapter is devoted to planning at several levels. It is sufficient to point out here that steps to integrate animals and plants into the agricultural pattern are often the result of direct planning, or the result of a newer technique. In the Australian wool-fat-lamb-wheat belt, subterranean clover species have brought about a revolution in mixed farming in recent decades. These clover varieties set seed underground, thus assuring regeneration when autumn rains follow. Sheep can be carried at a higher stocking rate and store lambs and sheep can be carried over into the next season if saleyard prices are too low, because the pasture has a greater degree of persistency.

There are other examples of private or official direction materially affecting plant-animal ratios or the substitution of improved varieties of livestock and crops. The same can be said of the introduction of an activity to the zone. In recent years, again in Eastern Australia, pork and bacon production (once 70 per cent confined to coastal dairy regions) has invaded the wheat-sheep belts. At first a sideline, today over 70 per cent of pig meat production is in these areas. Poultry—particularly turkey-raising—is another similar case. White veal production is a further example in the same belt, and beef is trending the same way. This changing emphasis in crop-livestock areas is in large part the result of direction from research.

Stages in diversification are easily recognisable, and become a major consideration, for probably no other form of agricultural activity lends itself to a study of sequent occupance as well as crop and livestock associations.

ii. Animals and Plants are Interdependent

The point raised here is the association of animal and crop. In the most intensive form of land use of this type, the crops are grown specifically to feed to animals. The advantage of this association to the commercial farmer is that the animal converts grain or other crops into protein with a higher market value. Maize-fattened cattle or pigs are the best examples.

Obviously this type of agriculture will be carried out in the more economically advanced societies. The subsistence or less developed economies cannot afford the luxury of raising animals. First, the availability of land is an acute problem. Even though animals can occupy relatively small areas when kept intensively, most parts of the monsoon world would suffer even if small amounts of land were taken over for livestock. Second is the time factor. Gestation periods involved in animal reproduction and the time involved in fattening young or store stock can be considerable. The balance of food sufficiency and famine is so fine that time is a major consideration. The rice growers of the world have so far found that improvements in strains of rice, the development of varieties with shorter maturing periods, and improved nursery and transplanting techniques have brought considerable success. It is unlikely, considering these facts, that animal husbandry can attract much attention. Thus, crop and livestock farming is essentially a commercial, 'Western' form of agriculture.

Even when the animal is the end product of the activity, considerable care is given to the production of grain and crops which provide the food for conversion. In the production of maize-fattened pork and bacon great effort is made to ensure the highest possible yields of maize. These are examples of interdependence on the farm itself—they may also exist on a world scale. Wheat and sheep in Australia have revealed this strongly in post war years. The heavy Japanese and European buying of Australian wool in the early 1950's created very high wool prices. Wheat production, though maintained at a high level, tended to be cut back in order to allow for the expansion of sheep flocks. A decade later the position was reversed. Heavy buying of Australian wheat and falling prices for wool indicated the close inverse correlation that exists.

Thus animals and plants have a physical interdependence on the farm unit and frequently relationships in marketing which may affect long term production plans.

iii. Production Methods Reflect Increasing Specialisation and Advanced Techniques

Degree of mechanisation is variable, but everywhere relatively high. This is only part of the high capital outlay on these types of farms. Fencing, barns where winters are severe, intensive shedding, haysheds, storage of water, silage pits and silos are some of the features of crop and livestock farming.

Use of fertilisers and pasture improvement is widespread. Superphosphate and organic manures too are spread by hand on the smaller farms. On larger farms inorganic manures are spread by motorised units or by aeroplane. The highly capitalised nature of this agriculture makes increasing intensification necessary. It also creates specialisation within the animal-plant association, so that one activity may become dominant on some farms to the exclusion of others, and groups of farms become dependent on other producers within the zone. This trend can be best summarised by considering a mixed farming 'zone' rather than a mixed 'farm'. The mixed nature of the agriculture is most obvious from observing a number of interdependent farms within an agricultural zone rather than a mixed farm considered as an isolated example.

There are many striking examples of improved techniques in both plant and animal genetic breeding which have permitted intensification of agriculture.

Coupled with the use of fertiliser, rotation, soil mechanics and so on previously outlined there have been other developments especially since the Second World War.

In the United States Corn Belt prior to the 1930's, only open-pollinated corn was grown. With sound farming techniques, high yields were obtained and many fine crossbred varieties were bred and introduced into the belt. With the advent of hybrid varieties—a seed initially produced by artificial hand pollination—the increased yield of maize has been staggering. Today over 500 million bushels more are produced on the same acreage as fifty years ago!

Vegetable hybrids have been produced, which not only have higher per acre yield, but frequently improved characteristics over the parent plants. These can include better colour, improved flavour, longer keeping qualities, earlier or later growing season and so on.

While wheat cannot yet be successfully hybridised to produce a commercial strain, the following outline of methods used in breeding of new wheat varieties illustrates the aims of the plant breeder and the skilled and patient techniques demanded.

A new wheat variety is produced by crossing two parent varieties which have between them the characteristics desired in a new variety. There may be, for example, a variety which has a strong straw and is highly productive. The strong straw is necessary to enable the crop to stand up to strong winds and also allow ease of harvest. The variety may have a low resistance to disease and have a low quality grain, however, and must be crossed with a variety which is both disease-resistant and has a desirable grain quality.

Grains are produced from the crossing of these varieties and are sown by hand in a special nursery plot. Seeds are harvested and sown under controlled conditions, so that individual plants can be observed and also harvested individually. Not all plants which result from such crossing will be satisfactory, and the great bulk of them will be discarded. The few plants which have desirable characteristics as outlined by the breeder will be retained for further investigation. The process is continued for a number of generations, with thousands of rows being sown, observed, harvested and checked for desirable qualities. Finally, comparisons are made by sowing the grains side by side with well known commercial strains. The yield of the various test grains is ascertained from the trial plots. Grain is then milled under experimental conditions and baking tests are undertaken.

The simplified procedures outlined above may take from ten to fifteen years and they involve slow and laborious techniques. The industry considers it well worth-while, for there is constant need for new varieties to meet changing conditions. The superior varieties so produced enable increased yields to be achieved, and frequently will result in superior varieties having other special qualities.

In like fashion, the animal industries have challenged geneticists. The chapter on dairying illustrates some of the techniques which have allowed huge gains to be made in quality and quantity of output when applied to this form of agriculture. New and improved breeds of beef cattle, sheep, pigs and poultry similarly reflect the increasing skill of scientists. The revolution in breeding, rearing, fattening and marketing techniques is seen by many world experts as one of the major counters to the world population 'explosion'.

In the last three or four decades both egg and poultry meat production have been revolutionised. From the free range extensive farming of a few hundred fowls has developed the modern egg factory where the fowl is virtually a machine. From fertile egg to grown fowl is a highly scientific and intensive process. Massive incubators hatching tens of thousands of chickens, supply the young stock. Their initial rearing, where fractions of cents are involved as the difference between profit and loss, creates complex procedures. Disease control, controlled feeding ration and housing are under constant scrutiny. The laying stock, housed in tens of thousands in the modern cage system farm, is subject to regulated artificial lighting, heating or cooling to produce optimum environmental conditions. Feeding, watering, egg collecting and manure removal are all automatically controlled.

Similarly the production of poultry for meat has undergone a revolution. Feed conversion figures have gradually become higher as the genetically bred strains convert plant protein to animal protein in a more efficient fashion.

While such techniques are striking in themselves, they highlight the final consideration in any study of modern agricultural production—that of marketing and distribution. This problem is greatest in the less developed economies where transport problems are greatest. In India the harvest for 1968 of rice, wheat and pulses is expected to total 102 million tons as against 89 million tons in the peak year of 1965, and only two-thirds of this in 1966 and 1967. Much of the credit for this will be due to the improved seed varieties bred for specific purposes, high yield being a major consideration.

A new hybrid rice seed, newer wheat varieties and the expanding use of hybrid maize strains are all playing their part. For example, the hybrid maize in southern India yielded double the padi rice tonnage on a per acre basis, yet used one-fifth of the amount of water. New rice varieties, developed with the co-operation of the Manilla Rice Institute of New Delhi and with aid from the Rockefeller and Ford Institutes, are among the biggest breakthroughs in developing countries. Yet despite these very promising results, and the future possibilities envisaged, the distribution of surplus is extremely difficult without an efficient transport system. Consumer-oriented products require well integrated marketing arrangements to ensure satisfactory disposal.

5. Other Considerations

Size of farms is variable. In the United States mixed farms approximate 120–200 acres generally, that is, about the size of farms in the temperate dairy lands. In Europe, however, they may be only half this size, while in the drier conditions of Australasia, South Africa and South America they are often over 1,000 acres. This great variation dictates what animals are kept and also the chief crops grown. The hog-corn belt of the United States specialises in pigs as the income earner, so crops grown are fed on the farm. The efficiency of pigs with a high conversion of grain to weight accounts for the small size of the farm unit. In temperate Europe much the same holds true, except that a variety of crops may be fed, plus dependence on some dairy cattle, often producing dairy cross beef vealers. Poultry may form an important part of farm routine.

In Australia maize is not grown widely, and wheat is the main grain fed to swine. The more usual association here and in South Africa is fat lambs, wheat and oats. These cereals may occupy one-fifth to one-quarter of the farm unit. The animals are grass fed and the flock of sheep may number from 500 to 1,500 or more. Size of farm will depend not only on the size of the flock, but also on the carrying capacity. This varies too but is generally in the range of one to three sheep per acre in the above areas.

Another consideration is the fluctuation in seasons, which causes considerable variation in the individual's returns and a consequent effect on national income. In a predominantly agricultural area or land, all sectors will feel the effects of variations in incomes. Further, there may be long term effects on investment in machinery and plant as well as replacement stock. Often the agriculturalist will wait for a series of above average years before undertaking major spending.

These considerations, based on the economics of production together with the limitations imposed by climatic and other natural factors, make it necessary for crop and livestock farmers to adopt the most efficient techniques possible. This form of agriculture is thus highly developed and constantly seeks to improve its methods. In these respects it is akin to advanced grassland farming and commercial dairying.

Bibliography

Alexander, J. W., *Economic Geography*, Prentice-Hall, Englewood Cliffs, N.J., 1963.

Buchanan, R. O., '*Some Reflections on Agricultural Geography*,' *Geography*, No. 44, 1959, 1–13.

Chisholm, M., *Rural Settlement and Land Use*, Hutchinson, London 1962.

Dumont, R., *Types of Rural Economy: Studies in World Agriculture*, Methuen, London 1957.

Higbee, E., *American Agriculture*, Wiley, New York 1958.

Thoman, R. S., *The Geography of Economic Activity*, McGraw-Hill, New York 1962.

Questions

1. 'Though established in European countries for many years, crop and livestock farming is comparatively new in North America, Australia and New Zealand.' Is this a reasonable assessment? Indicate why you have come to your conclusions.
2. 'In a world in which many countries are seeking new and expanding markets, the future of crop and livestock farming seems assured.' Critically discuss.
3. 'Diversification of activities in crop and livestock farming has many advantages and very few disadvantages.' Discuss this statement in relation to crop and livestock farming areas in two world areas.
4. 'The development of crop and livestock farming is likely to play an important role in the struggles of some nations for economic viability.' Discuss thoroughly, using specific examples where applicable.
5. 'The practice of crop and livestock farming in temperate lands allows the maximum utilisation of rural areas to be achieved.' How would you compare the potential of this type of rural activity with dairying?
State your conclusions based on factors as they apply in specific areas or countries.
6. 'Crop and livestock farming is an advanced farming activity characteristic of well developed countries. It envolves the use of the latest technological aids in achieving maximum productivity.' Critically discuss this statement with reference to specific areas.

13 Crop and Livestock Farming in South-Eastern Australia — the Wheat-Sheep Belt

1. Introduction

South-eastern Australia contains examples of crop and livestock associations in many parts of both the coastal valleys and plains, and in the interior riverine plains and surrounding areas. In the coastal lands the combination is of a variety of crops with dairying, while wheat and sheep are the major combinations inland. Also there are large areas given over to specialist land uses in the coastal areas—orcharding, viticulture, poultry and pig raising and market gardening, for example, while within the interior there is less variety, since drier conditions impose greater restrictions.

The main concern in this section is with the Wheat-Sheep belt, a region of fat lamb and wool sheep production in association with wheat and oat cropping, and with other activities to a lesser degree. Wheat and sheep are an association where both contribute to the farm income as well as contributing to the general effectiveness of farming, because of techniques common to both activities.

Although some farms receive the bulk of gross income from one activity—fat lamb raising, for example—the pattern of rotation of wheat and oats will also influence pasture establishment. This may also arise when decisions are being made as to what acreage of cereals is to be grown, and which parts of the farm are to be used for cropping. The establishment of long term leys and of permanent pasture is closely interwoven with cropping.

This has not always been the case. Earlier this century the production of wheat was largely monocultural, featuring long term cropping interspersed with simple fallowing. Today the establishment of pastures especially in the dominant fat lamb areas is often preceded by one, two, three or more years of continuous cropping. This gives the farmer the opportunity to break the soil down to a fine tilth, remove stones and stumps and partially eliminate such troublesome weeds as thistles. He will certainly apply additional quantities of phosphatic manures and often lime if warranted in the final season of cropping. Then the permanent pasture seed is sown together with the wheat or oats. After the cereal crop is harvested, a period of establishment is allowed for the grasses, which have been growing at ground level in the crop. The benefit of a thoroughly prepared seedbed for the pasture is obvious. Thus, although the final intention is a grazing paddock, crops may have paid for the establishment of the pasture in the preceding seasons.

In short term or semi-permanent pasture the new attitudes towards ley farming are also obvious in eastern Australia. Most farmers today sow inoculated clover seed as one of several grasses in the field. The inoculated seed ensures a high level of nitrogen-fixing bacteria in the root zone of the plants. Under a semi-permanent ley, a pasture may be retained for four or five years and grazed fairly heavily during this time. There is a considerable build-up in soil nitrogen during this period.

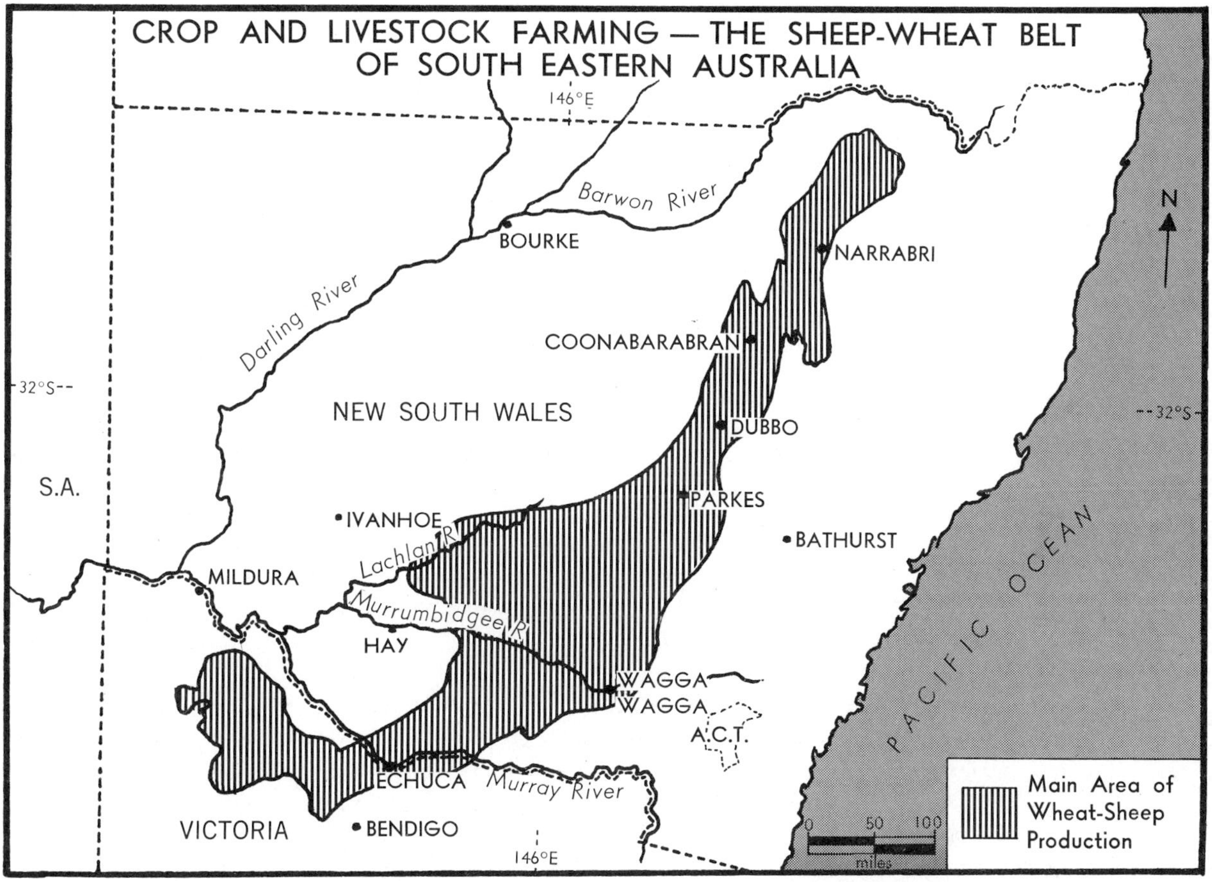

Obviously when the pasture gives way to a cereal crop there is considerable benefit to the cereal, and this is reflected in increased yield of grain. As there are many ramifications, involving all degrees of pasture permanency and the relative stress on grain, meat or wool, the relationships may become quite complex and involve sophisticated patterns of rotation. Obviously the close association of pasture-crop-animal is revealed.

The relationship may also be influenced by economic considerations. The simplest pattern is of extensions of either meat-wool or wheat depending largely on the strength of overseas demand. In the last few years Australian wheat sowings have increased, partly due to heavy demands on world wheat by the USSR, China and India. In some instances, too, beef cattle production has fluctuated, reflecting among other things changes in United States beef import policy. This too has affected the crop-stock relationships on some farms, but to a lesser degree since beef is proportionately less important in this belt. It is also worth noting that as many farmers in these areas have become increasingly 'fodder conscious', sowings of oats for hay or grain purposes have also expanded recently. The spectacular interest in improved pasture since the Second World War has encouraged grass seed production, notably of clovers and subterranean clovers in certain areas.

2. The Physical Background

The limits of the south-east Australian wheat-sheep belt are generally regarded as lying within the 10–30 inch annual isohyets, and a further generalisation is made that wheat-dominant farms tend towards the drier western margin while fat lambs become increasingly important in the more humid south-east. Again, crops occupy the open flatter areas on individual farms, while stock use the more sloping areas. It is also generally true throughout the whole belt that fat lambs are produced towards the eastern slopes and foothills and wheat in the western plains.

Recent alluvial soils of reasonable fertility are commonly found throughout the belt, although there are areas where variation occurs. Much more important to the land use pattern than variations in soils are urban influences, which create market-orientated dairying belts, for example. Orchards and market gardens may occur also, especially if irrigation water is available.

Natural vegetation throughout the region was primarily grassland with scattered forest. Grasses were short and well suited to the dry and hot summer conditions. These have been progressively replaced, except in the areas of upland grazing. The limited areas of forest are chiefly galerial in nature, and are found lining water-courses for a few hundred feet from the rivers and in lower lying areas around lagoons and billabongs.

i. Climate

The area experiences cool winters and warm to hot summers. Winter temperatures drop below freezing in the highlands and throughout the western plains where frosts are widespread, though snow is entirely absent on the plains. Heavy snowfalls are restricted to the highlands which lie outside the wheat-sheep belt. Killing frosts and out of season frosts may be a considerable risk to crops in the marginal west, especially in drier years. Unseasonal cold and windy spells are also possible and may create problems where early shearing is the practice. However, in no case are winters so severe as to create a need for permanent barns or shelter sheds; though windbreaks may be planted or natural shelter in the lee of hills utilised for newly-shorn flocks. As in all aspects of farming, careful husbandry is necessary.

The frost danger to crops can only be minimised by later sowings, but a later date hinges on rainfall at the optimum time too. The margin between early and late sowing is considerable. Most areas experience little difficulty in the initial sowing period, and quite frequently grain is sown in expectation of rain which is forecast in a few days time. What is vital is the distribution of rain throughout the growing period. In years when the expected rains do not eventuate, the decision as to whether to turn sheep into a sown field to eat a failing crop off or whether to wait for overdue rains, can become difficult. A crop paddock involves considerable outlay in terms of seed, fertiliser, fuel and labour.

While temperature and precipitation statistics give a good general background to the character of the region, they may disguise some of the problems and variations experienced. These lands well illustrate the dangers of interpreting climate statistics as being anything more than crude averages. A total annual rainfall mean disguises the distribution, which may have prevented the crop from being sown at all. Even monthly means may lose their significance if the rain was unusually concentrated or so light over a long period that it lacked effectiveness. Long successions of 'good years' may be interspersed with periods when conditions are below the norm. The arable acreage expands and contracts. Climate must be viewed as a synthesis of weather—an average of twenty inches of rainfall per annum means little—in a period of five years little more than half the average may fall, or it may be 50 per cent above the mean. From this one point firmly emerges the close link between economic planning in agriculture and natural climatic conditions and the extremes which may occur. In recent years a great deal of the success of farmers in the crop-

A flock of Merino sheep being mustered on Benangaroo, a sheep and cattle property near Coolac in the Southern Tablelands of New South Wales. In the background flows the Murrumbidgee River, with dark patches of irrigated lucerne on the river flats. The photograph clearly shows the more rugged margins of the sheep-wheat belt where cropping is limited by landforms. The galerial trees are red gum, box and low growing sheoak.

stock regions has been due to greater care in preparing for the seemingly inevitable droughts. It is unlikely that the dust bowl conditions of the early 1940's will ever be repeated, at least on the tragic scale of those years, in south-east Australia.

Wind and hail are other hazards. Better strawed varieties of wheat cope reasonably well with all but the strongest winds during the ripening period, and agronomists are continually seeking to improve this feature along with other desirable characteristics. British plant scientists who are well aware of storm damage to ripening fields envisage dwarf strawed varieties perhaps no taller than twenty inches. There is no protection against hail damage apart from crop insurance.

Oat crops, because of the loose nature of the grain head, are more likely to shake seed to the ground when ripe. This crop is grown for fodder in most cases and sheep soon clean up lost grain when let into the stubble. A final danger, though indirectly related to climate, is the risk of fire. In some years this has reached very dangerous proportions, and is a possibility every harvest. Fire insurance is often taken out as a means of protection.

Selected Statistics (*temperatures in degrees F.; precipitation in inches*)

Narrabri, N.S.W. (30° 20′S; 149° 50′E.)

	J	F	M	A	M	J	J	A	S	O	N	D	Year
Temp.	80	80	75	66	58	53	51	54	60	67	74	78	66
Precip.	2·8	2·9	2·4	1·5	1·9	2·3	1·8	1·6	1·5	2·0	2·4	2·4	25·5

Dubbo, N.S.W. (32° 15′S; 148° 37′E.)

Temp.	79	77	71	64	55	49	47	51	56	63	71	76	63
Precip.	2·1	1·9	1·8	1·9	1·9	1·9	2·0	1·8	1·9	1·6	1·8	2·0	22·6

Leeton, N.S.W. (33° 30′S; 146° 10′E.)

Temp.	76	76	71	61	55	49	47	51	56	62	68	74	62
Precip.	1·2	1·1	1·2	1·3	1·4	1·7	1·4	1·5	1·3	1·7	1·3	1·2	16·3

Tenterfield, N.S.W. (29° 03′S; 152° 01′E.)

Temp.	69	68	65	59	52	47	45	48	54	60	65	68	58
Precip.	4·3	3·3	3·1	1·8	1·7	2·3	2·3	1·8	2·1	2·7	3·1	3·7	32·2

ii. Topography, Soils and Drainage

Despite the extensive size of the wheat and sheep belt, there are remarkable similarities in topography, soils and drainage patterns over the area. The topography is generally level to undulating throughout, with rather more level land further west and undulating to hilly land in both the south and eastern margins towards the foothills of the Divide. Much of the Riverina is flat over very long distances, with occasional residual groups of low hills standing out 50–100 feet, generally not much more, above the level skyline. Such featureless topography does not give rise to local surface streams of any permanency, though the area is crossed by some important rivers with their catchment areas in the Great Divide.

The major streams include the Namoi, Macquarie and Castlereagh in the north; the Darling along the western edge; and the Lachlan and Murrumbidgee towards the south. All these streams are part of the Murray-Darling system and are important for supplies of stock and domestic water and for irrigation. No parts of the rivers are navigable, unfortunately, so that rail and road link the ports of Sydney and Melbourne with these riverine lands hundreds of miles from the coast. Between the coastal margins and the inland plains lie the plateau scarps and valleys of the Great Divide. This rugged and dissected upland has only limited areas of agricultural use and its western edge can be considered as the eastern margin of the belt.

Soils are based on light podsols and recent alluvials. There is a range in fertility, with the river silts being both most fertile and generally more satisfactory in water-holding capabilities. Here can be mentioned the problem of the lighter soils of the plains which dry out quickly under conditions of high evaporation.

All these soils react favourably to irrigation. The Murrumbidgee Irrigation Area soils are light sandy alluvials shading into heavier types in the river floodplains. Problems of saltation have been encountered in these soils, however, and water-logging in wet winters has occurred. In all areas potash is usually deficient and relatively heavy superphosphate applications are necessary. In recent years the bulk spreading of superphosphate by aeroplanes has become a feature.

3. The Nature of Sheep-Wheat Farming

i. Development of the Present Pattern

The historical pattern of this form of land use has several distinct phases. The earliest phase was a pioneer appraisal of the extent of the area in the beginning of the nineteenth century. Even in this exploratory period some false impressions of the nature of the landscape were engendered. Early explorers, who travelled much of the central western slopes and plains in attempting to settle the problem of the western rivers, viewed the area under conditions of both drought and above usual rainfall. As in the American midwest, early errors were made in attempting to farm this landscape with a heritage gained in the more sympathetic crop and stock lands of Britain and Europe, and the problem of drought was insufficiently guarded against.

The early development phase was characterised by some rather haphazard land granting and by widespread squatting. Water rights were alienated, better land was naturally chosen and poorer areas neglected. Many grants were too large for effective farming and grazing, and a few were too small. There was much careless clearing and burning of timber, the disastrous pattern of clearing that was established along the coast was repeated inland. Spectacular expansion occurred during the gold rushes of the 1850's and in the period of consolidation which followed. By the time the main gold fields were worked out the great bulk of the land had been permanently settled. This period was significant in establishing the basis of transport and settlement patterns, and the modern pattern is superimposed on the old in many instances.

The close relationship between wheat and sheep is a modern development, although sheep and wheat have been found side by side in the Riverina lands for over a century. The advent of rapid transport lines, allowing fat lambs to reach export ports as far as four hundred miles away, fragmentation of farms through cheap fencing materials, widespread use of subterranean clover, pasture improvement and superphosphate have all influenced the agricultural pattern. Economic pressures, the extra demand of the war years, the 1940–44 drought and increasing awareness of pasture, soil and water conservation have added their effects. Tax relief for capital investment, development loans and the growth of lending through the private sector have all affected long term decisions and given much-needed stability. Closer settlement and soldier settlement, some newer species of plants and improved animal strains have permitted an increased carrying capacity and are helping to overcome some of the effects of higher production costs.

ii. Capital Value and Farm Size

Land transactions of recent years have served to highlight the increasing value of land in this belt. Improved land will always fluctuate in price, depending not only on the inherent worth of the particular soil and its agricultural potential, but on capital improvements—buildings, both farm and domestic, yards, fencing, water facilities and so on—and seasonal conditions. In recent years the fluctuations have levelled out and the capital values of properties have become higher. Land values of over $100 per acre are commonplace, and they reflect the optimism concerning the capacity of the land to produce increased amounts.

Capital value of investment in machinery has been high in the area, allied with a very large increase in spending on grain and fodder storage facilities. The increased cost of labour has been offset to a degree by some of the developments such as bulk aerial spreading of superphosphate and crop spraying from the air.

Although the wool price boom years of the early 1950's can not be repeated, it is becoming more obvious that wool has successfully overcome the panic reaction to artificial fibres, and economic returns are relatively stable. Long term wheat contracts and expanding markets have levelled up prices for grain. Though internal problems of rising costs continue, the long term prospects seem favourable for meat, grain and wool.

There is a tendency towards reduction in the size of farms, at least when large estates come on the market; many have been fragmented for soldier settlement and closer settlement schemes. It is difficult to generalise on the size of properties, though 1000–1500 acres is often suggested as 'typical' of the better fat lamb areas, while twice this is probably necessary in wool-dominant regions. What is a better consideration is the carrying capacity—an expression of numbers of sheep per acre. As a consequence of this, the living area of the farm (the amount of land required to enable a family to make an adequate living) will vary. Thus if a flock of fat lamb dams of 600 is necessary together with 200 acres of crop and 400 additional reserve acres for implementing rotation (for 20 head of beef cattle, for land coming in to production, and so on), the living area will vary depending at the rate the sheep can be stocked. In the above example, 600 ewes stocked at one per acre need 600 acres, and only 450 acres if the carrying capacity of the land is one and a half sheep on the average. The size of the hypothetical farms in these cases will be 1,200 or 1,050 acres respectively—thus the living area.

iii. Crop and Stock Methods

Wheat is usually sown some time in early May, depending on prevailing conditions, and is normally preceded by oats. The crop paddocks may have been fallowed the previous spring, unless a new field is being brought into use after several seasons under pasture. There are no fixed ratios of grain to stock, and though 10–15 per cent of the farm's total acreage may be cropped this may be as much as half of the arable acreage available. Chemical spraying is frequently necessary in the first few weeks of growth to check thistles and other weeds. Some weeds like the deep rooted skeleton weed do not respond satisfactorily to weedicides and this consideration may result in permanent pasture. The crop paddock will usually have been deep ploughed, disc ploughed and harrowed to bring it into seedbed condition. Superphosphate is normally sown at the same time as the seed, separate boxes carrying seed and fertiliser.

The methods outlined are summarised as being representative of the whole area. There will be variations throughout the region as local conditions and individual decisions make some differences.

The oat crop is frequently lightly grazed by growing stock or in-lamb or recently lambed ewes, when it has made several inches of growth. The setback the crop receives is not vital to its eventual success when the grazing is light and in the early stage. Harvesting of oats is usual in late November or early December, while hay is cut about a fortnight before the grain is stripped. Modern hay balers using twine or wire ties, together with pick up and mechanical loading equipment, are speedy and

Woolclassers at work in a wool store in Geelong, Victoria, where individual bales are classed before being auctioned. This method of classing is adopted by graziers when the size of the clip does not warrant a wool classer being employed on the farm at shearing time. The woolclassers assess the relative fineness of the wool and express it in terms of the 'count'—a numerical value of the fibres.

A temporary method of wheat storage on the farm—wire mesh silos lined with hessian. The photograph is from Henty, Victoria. In the centre background an auger is transferring grain from the rear storage bin into a bulk bin on the truck. Note that the truck also has a grain auger mounted to handle grain from the harvester.

Permanent grain elevators store wheat while it awaits transport to its final destination. This shows the interior of a partly filled bulk wheat terminal at Geelong, Victoria. One of the most advanced storage sheds of its kind in the world, it was designed by engineers of the Grain Elevators Board. Its capacity is four million bushels of wheat, which is transferred by conveyor belts positioned under the centre of the roof.

Crossbred ewes on a wheat and fat lamb farm in the Australian Capital Territory being drenched for the treatment of the intestinal parasite liver fluke. A measured dose of drench is administered through the nozzle of the 'gun' held by the operator. The harness around his neck supports several pints of drench fluid, sufficient for many doses. Drenching is also carried out to combat a number of intestinal worms which infect the sheep.

efficient in handling the oblong bales. Steel haysheds have largely replaced the time- and energy-consuming hand stacked haystacks. If grain is to be stored it will probably be stripped and then placed in corrugated iron or concrete silos. It will be harvested by a modern combine harvester and loaded out of a bulk bin on a truck and into the silo by motor driven auger. Some seed may be retained for the following year's sowing, some may be sold.

Wheat varieties vary from hard milling wheats generally in the heavier soils of the north of the zone to lighter softer varieties in the south. But there are many varieties, some of recent origin, some that are well tried favourites. Newer varieties are often introduced through the work of Department of Agriculture officers, as the result of test plots sown and appraised in farming journals or in farm broadcasts. The suitability of wheat varieties to different areas varies, but there is an increasing awareness that certain varieties should be sown in different areas.

Sheep varieties are usually classified as mutton or wool types, though this is an oversimplification. As far as purebred wool varieties are concerned the Merino is the only major breed. Two moderately popular dual purpose types are the New Zealand and Australian developed Corriedale breed, and rather less in importance, except in the cooler and wetter south, the Polwarth. The remaining group of British breeds are seldom kept as a purebred proposition except on stud farms, and these are largely mutton types—the Romney Marsh, Southdown and Border Leicester, and the Dorset Horn or Polled Dorset.

The Merino is found throughout the western zone where wool production is dominant, and within the zone where special conditions exist, for example in hilly areas of coarser grass where extensive grazing is carried out and in undeveloped areas such as forest leases. Elsewhere the ratio of purebred Merinos to other breeds is low.

Merino dams are commonly used, however, to produce first cross progeny. The sire used is a British breed, commonly the Border Leicester. Often other crossbred lambs are produced using Dorset Horn or Polled Dorset sires; occasionally Romney Marsh and Southdown are employed with Merinos. The Corriedale is sometimes used, but is more often kept as a dual purpose purebred.

First cross ewes like those mentioned above are often retained as breeding stock. Where conditions suit, the first cross progeny are marketed as fat lambs. In the wetter areas, and particularly in irrigation districts, breeding is not widespread and stock are bought from the drier districts. Particularly favoured are the first cross ewes which are again mated to a British breed—commonly Dorset Horn, Polled Dorset, Border Leicester and Southdown. The resulting lamb is a second cross, carrying two crosses of mutton type and one of wool type in its makeup. The advantage of this method is that the resulting lambs are a better meat proposition, and the ewes produce a fine comeback wool which has brought satisfactory prices over the years. Such large-framed ewes as are produced by a first cross are capable of giving a high milk supply, allowing lambs to be marketed as sucker lambs. These lambs bring premium prices for both local and overseas trade.

Shearing in progress at Wanbanumba, a grazing property near Young on the south-east slopes of the New South Wales wheat-sheep belt. Here the 'picker-up' kneels to lift the fleece shorn in one piece. He will throw it onto the skirting and classing table, where the rough soiled edges of the fleece are picked off. The fleece is then baled with others of similar type.

If lambs have to be retained because of seasonal conditions, then they can be shorn and produce a fine quality wool. Allied with the production of lambs are also the stud properties which produce high quality breeding stock. There is considerable pressure to maintain and improve the quality of bloodstock in both purebred and crossbred flocks.

iv. Other Activities

Although wheat and sheep are by far the most important activities of the zone, there are considerable amounts of other agricultural products marketed from this region.

a. Cattle—dairy and beef. It has been indicated that dairy cattle will usually be found where urban populations create a market. Around most of the larger towns of the wheat-sheep area, local dairies exist. Conditions are not necessarily completely favourable—rainfall is generally too low, though limited irrigation facilities may be present, and summer and winter temperatures are extreme. The extremes of temperature do not directly affect the cows, but limit the growth and level of nutrition of grasses. In such cases the herd production will bear little comparison with the favoured dairy zones. Nevertheless, economic returns are usually very satisfactory, in view of high prices for whole milk and cream, and a restriction of production in line with the size of the local market.

Surplus milk can be directed towards pigs if seasonal conditions produce a flush. Larger towns have cream and butter factories supplying local demand and in many instances have expanded to produce bacon and pork smallgoods. Because there are likely to be periods of relatively high milk production, in order that the midwinter and midsummer quotas be met, calf production is also a feature on these dairies, as it is in a small way on most farms in the better watered areas.

Further out from the townships beef cattle herds increase and dairy cows are kept only for the farmer's own needs. There is no real distribution pattern of beef cattle, though areas of greater slope, timbered regions and less developed leasehold land will have greater numbers. There are trends towards increasing numbers of beef cattle, based on continuing high prices in both local and overseas markets. The 1967–68 foot and mouth disease outbreak in the United Kingdom also gave a temporary lift to the export market for beef in particular.

There is a preference for Hereford cattle and Hereford cross throughout the area, with Beef Shorthorn, Aberdeen Angus and Red and Black Poll breeds and crosses also popular. Dairy crosses are usually kept rather than purebred dairy cows, since they cope better with the hotter and drier conditions.

b. Pig raising. There has been a marked swing from the dairy areas to the wheat growing areas in the production of pork and bacon pigs in eastern Australia. Today the wheat-sheep belt is responsible for over two-thirds of all production. There has been a considerable revolution in methods of production accompanying the change. Modern techniques are rapidly replacing the haphazard sideline attitudes widespread several decades ago.

In part these changes have helped to create more stable prices for pig meat, and in turn the establishment of farms chiefly dependent on pigs for the bulk of farm income have been encouraged. Market demand has increased as changes in taste, largely influenced by European migration, are introduced. Accompanying this is pressure on retail prices. European and British migrants consider pork as much a part of their diet as beef and lamb in Australia and are not prepared to pay the high prices of the past. There is increasing demand and pressure to reduce price.

Three factors are helping to establish lower production costs: the establishment of more economic production units, better techniques (particularly in feeding) and new or better strains and breeds of pigs. The establishment of modern piggeries is a welcome change in the industry, since for many years pigs were often housed in second rate buildings and fed poorly balanced foodstuffs. Better building materials, automatic and semi-automatic watering, more efficient designing—especially in farrowing pens and where young stock are concerned—are all features of the newer units and are being introduced on a small scale on some farms. Of particular importance is the scientific preparation of feeding rations, replacing the haphazard mixtures typical some years ago. There is awareness that balanced nutrition is most necessary to produce a well finished animal with the correct ratio of fat to lean meat, and also if economic costs are to be lowered and profits raised. Many wheat and sheep farms are becoming interested in small, well managed units as an adjunct to their main activities.

The other major influence has been the impetus given by the introduction of Danish and Swedish Landrace pigs. This breed has a high feed to meat conversion rate, is quick maturing and well conformed to meet market requirements. It is used in large numbers as the basis of crossbred combinations. Not the least important influence has been a quickening of interest in improving the strains of the established breeds, particularly the Berkshire, Tamworth and Large White pigs.

c. Minor activities. Clover seed production, poultry raising, beekeeping and fruit growing are minor activities in respect of the areas involved, though they may be highly significant within certain areas. Pasture seed production has increased greatly in response to demands for improved pasture. Although clover is the main variety of grass on a large scale, there is growing interest in all species such as lucerne, various rye grasses, phalaris and others. Specialist areas of poultry production are arising as demand changes—for turkeys, for example. Apiaries are restricted to the eucalypt forests around major rivers or in hill areas. Some specialist areas of orcharding exist.

Bibliography

Andrews, J., 'Pioneer Settlement: an attempt at synthesis,' *Austr. Geographer*, 1, 1932, pp. 18–34.

Atlas of Australian Resources, Canberra, Department of National Development.

Meinig, P. W., 'Colonization of wheat lands: Some Australian and American comparisons,' *Australian Geographer*, 7, 1959, pp. 205–13.

Rutherford, Logan and Missen, *New Viewpoints in Economic Geography*, Martindale Press, Sydney 1966.

Year Book of the Commonwealth of Australia, Bureau of Census and Statistics, Canberra (annual).

Questions

1. 'The crop-livestock association found in modern Western agriculture is complementary as well as supplementary.' Discuss this statement with reference to the wheat-sheep area of south-eastern Australia.
2. 'The sequent occupance of south-eastern Australia in regard to crop and livestock farming areas was and is largely predictable, particularly in the light of its relatively recent settlement.' Do you agree with this statement ?
3. What evidence is there of diversification in crop and livestock farming in south-eastern Australia ?
4. 'The limitations to the various farming activities in south-eastern Australia are a combination of both physical and cultural considerations.' Carefully indicate the spatial and areal differentiation of agricultural areas in New South Wales.
5. Write an essay with the title 'Technical change and its impact on agriculture'. In your essay you should pay particular attention to Australia and New Zealand.

14 Mixed Crop and Livestock Farming in the Central Canterbury Plains

1. Introduction

The Canterbury Plains forms the largest area of arable land in New Zealand. It consists of an area of flat to undulating plains, coastal swamps and low hill and valley land. Its generally low elevation and gentle slopes, combined with a moderately wet and cool temperate climate make it an area admirably suited to cereal growing and the production of a wide range of field crops. In association with this variety of crops, the region grazes 12 per cent of all New Zealand's sheep (although less than 3 per cent of her cattle), and thus presents an excellent example of mixed crop and livestock agriculture.

Although the area of the Canterbury Plains represents only 7 per cent of New Zealand's occupied farmland, this narrow strip averaging about 40 miles wide and only 250 miles long produces 40 per cent of all field crops, and 65 per cent of all grain crops grown in the Dominion. The importance of the plains as a grain cropping region can be gauged from the fact that it grows 75 per cent of all wheat and barley, and 50 per cent of all oats grown in the country.

A remarkable variety of agricultural activities characterises the Canterbury Plains. As well as the large amounts of wheat, oats and barley, substantial quantities of maize, lucerne, potatoes and peas are grown. Lucerne, clover and grass seed production is important in the east of the belt. Turnips, chou mollier, swedes, kale, ryecorn, fodder beet, lupins and rape are among the wide variety of fodder crops commonly grown. In North Canterbury some apples and significant quantities of market garden crops are grown near Loburn, while in the South Christchurch outer suburbs, market gardens and some orchards are found. Special crops like export strawberries are grown at Timaru, and glasshouse culture and market gardens are found in the Port Hills district. Where urban populations create local markets, town supply dairies for whole milk are found. On the outskirts of Christchurch, piggeries and poultry farms as well as a number of dairies are located in response to the city's needs. In the less favourable landform zones of the western margins, and in tussock grass zones, some wool sheep and beef cattle are reared.

Despite the importance of these varied activities, the wealth of Canterbury is based on fat lambs and cereal crops. These two activities in turn closely influence the rotation pattern and the establishment of permanent improved pasture. The period of short term pasture will also be influenced by the stress on fat lambs or wheat. If wheat acreages are to be expanded—which conversely precipitates a cut-back in sheep—then pastures are ploughed up to become wheat fields. Seldom will changes in the crop-stock ratio be solely the individual farmer's choice—the likely influences are economic considerations of the market. In New Zealand's case this means the British meat and dairy market, thousands of miles from the centre of production.

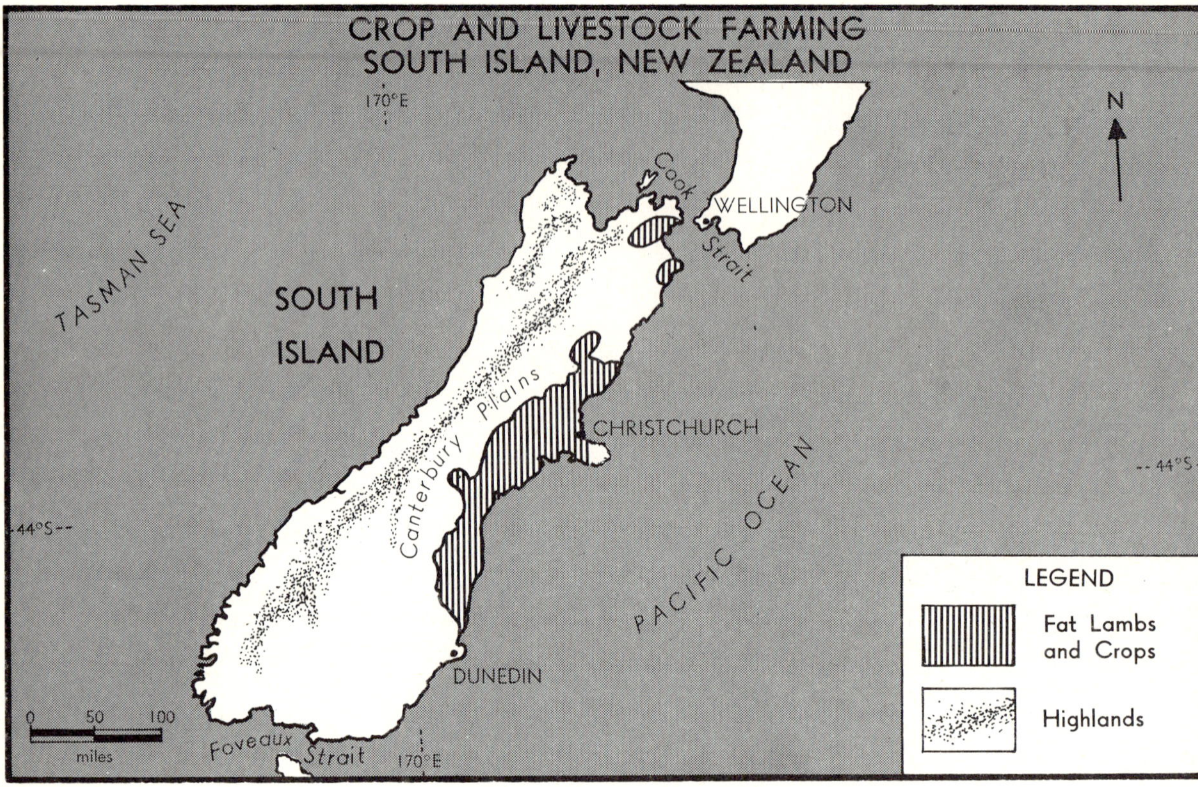

2. Coastal Plains and Western Downlands

Although the Canterbury Plains is frequently regarded as a single topographic unit, there are in fact variations in landforms throughout the region, and differences in structure and fertility of soils. These physical variations affect farm size, carrying capacity of stock, the types of crops grown and rotation patterns employed. The division studied as representative of mixed crop and livestock farming is the Central Plains. On either side of this region are smaller distinct areas which are considered briefly.

The coastal plains lie in a very narrow fringe on three sides of Christchurch, and on other parts of coastal lowland. Of heavy loam and silt soils these areas are well suited to the establishment of long term pasture used for cattle fattening and dairying. These activities are seldom seen elsewhere in Canterbury. The swampy lowlands have been extensively drained to produce fertile soils having considerable water retaining properties. Dairying is concentrated in the districts of Belfast, Halswell and Taitapu; in addition, there are over two thousand acres devoted to intensive garden farming. Included here are market garden crops of many kinds—peas, potatoes, celery, carrots, onions, beetroot and beans; special crops such as asparagus, tomatoes, strawberries and flowers are also important. As well as these vegetable and fruit crops, the coastal plains produce clover seed, livestock fodder crops and wheat, oats and barley.

The western downlands abutt the central highlands, and are a discontinuous belt interspersed with hill and valley country forming the western boundary of the plains. The factor which decides the inclusion of sectors of country as part of the Canterbury Plains is the suitability of the land for arable farming. If slopes are ploughable, then satisfactory crops of wheat, oats and barley can be grown under a climate regime which experiences reliable ripening weather and well distributed and adequate rainfall. Much of this division has a lower natural fertility than the remainder of the plains, although the clay content of the soils is high and water retaining properties better than in soils of the adjoining central areas. These downland soils of lower fertility require considerable additions of phosphatic manures to overcome deficiencies and lime to correct acidity. Many farms in this belt have achieved considerable gains in soil fertility by a combination of superphosphate topdressing and judicious pasture management. Although pastures do not have the all year round high level of nutrition of the eastern areas, they nevertheless produce considerable numbers of export quality fat lambs.

3. The Central Plains

i. Mixed Crop and Livestock Characteristics

The Central Plains is by far the largest division within the Canterbury Plains, and lies between the coastal plains and the western downlands. It is primarily a dry cropping and grazing zone, with wheat the major grain and crossbred sheep the main stock. Both mutton and lamb are produced from this zone, although the main intention is to turn off sucker lambs and hoggets for overseas markets. Wool from the half-bred or threequarters-bred ewes is also important, and a wide range of

High carrying capacity country (five to seven sheep to the acre) is typical of much of the Canterbury Plains. The river terraces in the background are clearly visible, as is the course of the river beyond the middle distance. Such rich river flats are highly pasture-improved and also produce heavy yields of fodder crops.

green feed and other fodder crops, which help maintain the high carrying capacities typical of these Canterbury farms. But the basis of farming activity is grassland, for field crops usually occupy less than one third of all farm land. Because of the dependence on advanced grassland techniques, rotation patterns involving pasture are vital to the success of any long term planning. The need to maintain soil fertility, when production of over a hundredweight of wool and meat per acre is annually achieved, is obvious. There are continual developments in better fertilisers, the introduction of higher yielding grasses, more efficient grazing techniques, changing preferences in fodder crops and consequent increases in the rates at which farms are stocked.

ii. Size of Farms and Carrying Capacity

Farm size is usually between 300 and 500 acres in the plains, and internal divisions will create ten to twenty fields, often laid out in a symmetrical pattern. The size of these fields may be as small as five acres, and seldom over forty acres. Carrying capacity on these farms is usually between three and four ewe equivalents per acre. The ewe equivalent figure is obtained by giving a numerical value to all classes of livestock as follows:

Ewes—value, 1; Hoggets—value, ½; All other sheep—value, 1; Cattle—value, 5.

Thus on a farm of 551 acres carrying 1,370 ewes, 538 hoggets, 16 rams and 30 head of cattle, the total ewe equivalent figure will be 1,805. To obtain the carrying capacity, the figure of 1,805 is divided by the acreage of 551, giving 3·28 ewe equivalents per acre. The carrying capacity figure will be higher if the cash crop acreages are deducted from the total acreage of the farm.

iii. The Background of the Present Pattern

The area has not always had a mixed farming character. It was first settled in the 1850's, and the early farmers built up extensive woolgrowing farms based on Merino sheep, that ranged from 20,000 to 50,000 acres or more in size. Much of the area was covered with tussock grasses and tea-tree (manuka) scrub. Clearing was thus comparatively easy. Paddocks were periodically burned to encourage regrowth, and this was a factor in the beginnings of widespread soil erosion, as winds blowing from the central highlands attacked the unvegetated fields. This era of farming was based on extensive methods and little effort was made to improve pastures or conserve fodder. As a result, there were serious difficulties when shortages of feed occurred in winter and at the end of summer.

Over the next twenty years wheat was introduced into the plains as a major cash crop. The growing of wheat quickly became widespread and many farms adopted a monocultural approach. This was a retrograde period as far as the maintenance of grass cover was concerned, and the problems of soil erosion became serious on many farms. The overgrazing and overcropping were especially serious when followed by below average rainfall. Some benefits did accrue from the swing to wheat growing, however, for fencing and subdivision became necessary. Where wheat and sheep were found on the one farm the basis at least of some simple form of rotation was laid.

A new era was heralded in 1882, with the introduction of refrigerated holds in overseas cargo chips. From this time onwards the character of the plains changed, as increasing demands were made for 'Canterbury lamb' in the markets of England. As mixed farming became commonplace, woolgrowing shifted towards the highlands and into the open expanses of the south. Fat lamb production and cereal

Mustering Romney cross ewes and young lambs on Hiwira Station, in the Rangitikei district. The mixed nature of the farms, which incorporate fat lamb raising with beef, is typical in the areas on the fringe of the higher country. Although there is evidence of gully erosion on the slopes to the left of the photograph, most of the higher hills are well covered with mixed forest. Both Black Poll and Hereford cattle are visible.

cropping became the cash activities, and newer pasture grasses and fodder crops were added to create the pattern of today's mixed farms.

iv. Topography and Soils

The Canterbury Plains resulted from the coalescing fans of the Waimakiriri, Rakaia and Rangitata rivers and their tributaries. Thus over much of the region the landforms are extensive flat to gently undulating plains with an eastward slope. At a distance of thirty miles from the coast the rivers flow around two hundred feet below the level of the plains. The gently sloping and nearly featureless plains of the central region are ideally suited to the widespread use of machinery, which is employed in all aspects of arable farming.

Although the soils close to the river valleys are mainly alluvial, much of the Central Plains consists of loess covered shingle deposits. Local names frequently differentiate soils, mainly on the basis of depth of topsoil. The soil type occupying the largest area of arable plainland is the Lismore type. This consists of a 4 to 6 inch light grey-brown loam over about twice this depth of lighter grey-yellow silt. These light soils overlay a coarse gravel and stone layer, although the topsoil is relatively free of stones. A shallower variation of this type is the Eyre shallow silt loam, with a topsoil about 3 inches deep, but again fairly free of stones. Rather deeper than the Lismore is the Chertsey silt. This yellow-grey type has a topsoil about 8 inches deep, lying on some 10 inches of light brown-yellow silt loam on stones. The topsoil is relatively stone free.

v. Climate

Over much of the Central Plains the rainfall average is from 20 to 35 inches per annum, which in these temperate latitudes seems sufficient from both a cropping and pasture viewpoint. However, these averages disguise the harmful recurring droughts which are an established threat, even though such droughts may usually be considered moderate in severity compared with those in other regions of the mixed farming world. There is also a degree of variability in rainfall both from year to year and month to month, and drying north-west summer winds which considerably reduce rainfall effectiveness. In winter, too, strong cold south-westerlies dry out pastures and erode topsoil in newly ploughed fields. Windbreaks and shelter belts reduce erosion effects considerably on the open plains. The summer rainfall is reduced in its effectiveness by high evapotranspiration rates, and it is estimated that three to five months of the year may have soil moisture below the figure of 25 per cent which is considered necessary to maintain pastures adequately.

Selected Statistics (temperatures in degrees F.; precipitation in inches)
Christchurch (43° 30′S; 172° 35′ E.)

	J	F	M	A	M	J	J	A	S	O	N	D	Year
Temp.	61	61	58	54	48	43	42	44	49	53	56	60	53
Precip.	2·2	1·8	1·9	1·8	2·8	2·6	2·6	2·1	2·1	2·0	1·8	2·5	26·2

Export fat lamb carcasses being checked before being bagged and shipped out from the Hawkes Bay Farmers Meat Works, Whakatu. The uniformly high quality of New Zealand lamb is a well established fact in British markets, and meat destined for overseas is carefully graded.

Monthly temperature averages are around 40°F. in winter to about 60°F. in summer, and there are usually four or five months of the year below 50°F. Thus conditions are mild to warm, lacking the great extremes of temperatures that characterize the interior livestock and crop belts of the large continents. The dangers of sudden cold weather after shearing or during lambing must be guarded against by the planting of shelter lines of trees, usually *Pinus radiata.* Gorse hedges are also often planted between fields to serve as windbreaks. Light snowfalls are possible during winter, and when these are followed by strong winds conditions can be dangerously bleak.

vi. Livestock Management

Since fat lamb raising is the major income earner on these farms, great attention is paid to maintaining a flock of breeding ewes capable of producing a high percentage of well-framed progeny. In many instances the ewes are bought from breeding farms in the highlands and the tussock country. They may be bought as lambs, hoggets or as older sheep—two or three year olds. Rather than replace a breeding flock completely when the ewes are aged, most farms prefer to cull a third or so of the flock in each of three consecutive years, for example. The culled ewes are fattened and sold, and the remaining handpicked ewes are retained in the flock, being considered sound enough for a further season or two of breeding. Enough new breeding ewes are then bought to keep the numbers at the required level. Some farms prefer to breed their own replacements, and usually will replace a percentage of the flock annually.

Several breeds and crosses are used in fat lamb production. On some of the larger farms in drier and more rugged areas, a Merino crossbred is favoured. These ewes will be mated with a British breed to produce a second cross lamb. By far the most popular breed, both as a purebred and as the basis for crossbreeding to produce a fat lamb breeding ewe, is the Romney Marsh. The heavy, square framed Romney sheep have been found ideal for producing a well conformed lamb. If used as a purebred dam, the Romney Marsh will be crossed with a British breed, frequently the Southdown. If Romney cross ewes are dams, then the sire will usually be Romney Marsh, Southdown, Dorset Horn or Polled Dorset, Dorset Down or Border Leicester. The New Zealand-evolved Corriedale sheep is a popular breed, both as a dual purpose mutton-wool prospect or as the basis for crossbreeding.

Some farms run a small stud to supply their own flock breeding rams, and to produce a few rams for sale. On average sized farms the flock is usually under a hundred, and the breed will probably be either Romney Marsh, Dorset Horn or Southdown. From these small stud flocks enough ram lambs are bred to maintain a ratio of one ram to 60 or 80 ewe breeders in the fat lamb flock.

Of great importance on these intensive farms is the percentage of lambs produced. With cross-bred ewes capable of producing and rearing twin lambs in about one third of all matings, lambing percentages exceed 130 per cent in some cases and averages of 110 per cent are commonplace. Lambs are trucked off the farm in small equally graded batches of usually less than a hundred at a time, most of them being marketed from November to January. Some smaller numbers may be sold later in the autumn, and a few may be carried over and sold as hoggets in the spring. Saleyard weights average around 35 pounds for these lambs, so that meat output per acre on these farms may be upwards of 100 pounds; in addition is the 25 to 35 pounds of wool commonly produced per acre.

vii. Fodder and Cash Cropping

A wide variety of crops is grown for fattening stock in the field, and harvested for hay, silage or seed. Popular crops grown as greenfeed include rape, barley, oats, wheat, ryecorn and sweet lupins. These may also be cut for hay or silage in later stages of growth, although lucerne is by far the main hay crop grown. Other fodder crops that are sometimes grown in areas of better rainfall, but more usually irrigated from rivers or ground water, include fodder beet, chou mollier, turnips, swedes and kale. Yields of these crops are often very high considering only two or three complete irrigations are used in a normal year. Yields for swedes in mid-Canterbury range from 15 to 30 tons per acre, and commonly average around 20 tons. In a random example in this area a twelve acre plot of irrigated swedes yielded an estimated 24 tons per acre. This was sufficient to carry 1,100 ewes and 500 hoggets for a period of twenty-two days, and 340 wethers for a period of forty-seven days. All grazing took place between 3 July and 20 August, indicating the value of this crop as winter feed. The only additional feed was one bale of hay per hundred sheep per day.

Potatoes and peas are major cash crops on many farms, and although sown about the same time, they play a different role in the rotation scheme. Potatoes usually precede permanent pasture, since the deep working and heavy application of fertiliser needed to produce a high yielding crop makes an ideal seedbed for grasses after the potatoes are dug. Peas are a leguminous and leafy crop well suited to a short rotation, since after harvest the residue of the crop is ploughed under, adding both nitrogen and organic matter to the soil. Thus wheat is an ideal crop to follow peas, and it may also follow rape, lupins and red clover.

Wheat is not sown in the same field every year normally, but possibly twice in four years in the same field, interspersed with legumes. It has also become incorporated into longer term rotations, where there is less running down of soil fertility and less damage to topsoil structure from repeated ploughing. In the course of events which brought about increased use of superphosphates and leguminous pastures, the yield of wheat has increased on a per acre basis. Many farms that were producing less than 30 bushels per acre averages at the end of the Second World War, now consistently average over 60 bushels per acre. Often the selected varieties, which include Cross 7, Gabo, Aotea and Arawa, yield up to 90 bushels per acre under suitable conditions.

viii. Pasture Management

Despite the importance of cash and fodder crops, the Central Plains farms devote more than one half of their acreage to pasture. In recent years there has been a considerable change in attitudes towards the establishment of permanent and short term pasture. The light soils of the plains dry out quickly in years of below average rainfall, and with overcropping become dusty and easily eroded by wind. The realisation that overcropping had caused a marked fall in soil fertility and had affected soil stability became very obvious after the Second World War. As a result many farmers turned to a pattern incorporating long term pasture rotation and an expansion of fat lamb production. The buoyant wool prices of the fifties was a further incentive to increase pasture and restrict wheat acreages.

Subterranean clover has pioneered a great deal of the pasture development in the Canterbury Plains. The ability of the species to regenerate from sub-surface seeds following even relatively severe drought meant that a substantial clover-to-pasture basis was established. On most farms, a few seasons after subterranean clover was

Romney Marsh and Southdown cross hoggets graze semi-improved pastures in the Tapanui district on the edge of the Blue Mountains. On the plains, the improved pasture and crop paddocks contrast with the scarcely used slopes. Gorse hedge and pine windbreaks are found along many fence lines.

Aerial topdressing is one of the best examples of these modern techniques in New Zealand's pastoral industries. Here a Fletcher topdressing aircraft drops superphosphate in a wide swathe on hill country inaccessible to conventional spreaders. The loading machine in the central foreground fills the aircraft's hopper rapidly, to allow for a speedy take-off after each run.

established in some fields, it had become a volunteer in all fields. Sufficient seed had been transferred by stock from other fields to make it unnecessary to sow it in new pastures.

Other grasses commonly used in long term pasture mixtures include perennial ryegrass, dogstail, white clover, Montgomery red clover, broad red clover, cowgrass, cocksfoot, timothy and lucerne. Although some of these grasses are also used in short term rotation pastures, Italian ryegrass or some other annual replaces the perennial variety, and suckling clover and cereals are also frequently used.

Both phosphates and sulphur are commonly deficient in the Canterbury light soils. Superphosphate corrects both shortages, and is spread from the air when large acreages are involved, otherwise mechanically from the ground. The amount per acre usually varies from one to two hundredweight per acre at the time of establishing the pasture. A similar amount may be used annually or biennially as topdressing. The use of DDT superphosphate, especially before selected paddocks are closed for winter feed saving, is a safeguard against the pasture pests, porina and grass grub. The addition of lime does not directly affect soil fertility, but these soils are lime-deficient and from one half to one ton per acre is a usual application when establishing pastures. The use of lime is to correct acidity imbalance—the 'pH ratio'.

The future of crop and livestock farming in the Central Plains demands increasing productivity and stabilisation of costs where possible. Modern developments such as hormone control of breeding cycles in ewes (allowing the period of lambing to be telescoped into a short period) are among aids available to assist in the programming of the farming calendar. Improved body weights in ewes, better species of crops and pasture grasses and the introduction of new varieties of fodder all play their part. The continued success of Canterbury's agriculture will depend on successful market techniques adopted to gain a share of the United Kingdom's £374 stg. million meat bill. In 1966, when the import of foreign meat reached this level, New Zealand had a £71·6 million share—about 20 per cent of the market. Britain's possible entry into the EEC is a threat to New Zealand's agriculture which makes new appraisals of export markets of great concern. For the immediate present, high quality New Zealand meat is well regarded in British markets, and this clearly reflects the efficiency of the mixed crop and livestock farming of the Central Canterbury Plains.

Bibliography

Levy, E. B., *Grasslands of New Zealand*, Government Printer, Wellington 1955.
Rutherford et al, *New Viewpoints in Economic Geography*, Martindale Press, Sydney 1966.
Cumberland, K. B., *New Zealand Topical Geographies: Agriculture*, Whitcombe and Tombs Ltd, 1967.
Coop, I. E., *The Principles and Practice of Animal Nutrition*, Government Printer, Wellington 1961.
Curry, L., 'Regional variation in seasonal programming of livestock farms in New Zealand,' *Economic Geography*, 39, 1963, pp. 95–118.
New Zealand Journal of Agriculture (monthly) for research articles.
New Zealand Official Yearbook, 1964 and later.

Questions

1. What evidence is there of diversification in New Zealand crop and livestock farming?
2. Account for the fact that the Canterbury Plains is the 'granary' of New Zealand.
3. Discuss the methods of livestock management, fodder and cash cropping and pasture improvement in the Canterbury Plains.
4. How does the land use pattern of the Canterbury Plains bring out the classical theories of agricultural location and utilisation?
5. Discuss the implications for the Canterbury Plains if Britain:
(a) were to join the EEC
(b) were to ban the import of meat from Denmark into Britain.
6. In what ways do the crop and livestock activities in south-eastern Australia and New Zealand differ?
7. Why are New Zealand's meat exports to the United Kingdom more than double those of Australia? (1966 figures in £S millions: New Zealand = 71·6; Australia = 32·8.)

Part Six

The Geographical Impact of Planning

15 The Geographical Impact of Planning

Introduction

The individual farmer, and thus the whole pattern of agricultural land use, works within a broad framework of physical and cultural factors. These physical factors include considerations of topography, soils and climate. The cultural factors include the state of the economy, traditional and political forces.

It is only rarely that geography delves into politics, but where political decisions greatly influence the pattern and function of land use then it is highly relevant. In the case of agriculture, there are few areas on earth where government action has not been greatly or partly responsible for its structure. This is due to the peculiar nature of agriculture throughout the world, whereby it is as much a way of life as it is a form of economic activity.

This government action may occur in all types of economies and may be in a variety of forms for a variety of reasons.

It must also be recognised that government action may be conducted on various levels from international arrangements to local convenants. In every case, however, there must be underlying reasons, some of which are misguided but nevertheless usually exist for a specific economic objective.

1. The Objectives of Planning

Planning at any scale is aimed at influencing a change in agricultural land use patterns. Although these reasons and patterns may be closely interrelated, they can be listed under the following headings:

i. Development of new lands; for example, the Virgin Lands Scheme of the USSR.
ii. Rehabilitation of adversely affected lands; flood and drought are good examples.
iii. Policies of correcting maladjusted settlement; for example, uneconomic farm sizes leading to pasture deterioration.
iv. Resettlement opportunities; for example, Soldier Settlement, often involving sub-division.
v. Changes resulting from research; for example, new or improved crop strains and trace element application.
vi. Economic sanctions and restrictions due to supply and demand of specific commodities; for example, sugar acreage largely governed by international agreements.

The above considerations do not include all the possible combinations of planning, particularly when it is remembered that agriculture is often a way of life and therefore subject to many non-economic and indeed non-rational forces. However, in isolation, they oversimplify the concepts—in combination they reveal the intricacies and complexities of attempting to vary an existing pattern of land use—in the way of changing processes and sequent occupance.

It must be noted also that there is often a gap between recognition and desirability to actual implementation and effect. These considerations, working singly or in combination, are manifest in many forms, dependent on the type of government and state of economy. A consideration of these factors is discussed further in this chapter.

i. Development of New Lands

Few parts of the earth's surface these days can be categorised as 'new'—perhaps the term 'little-used' is more appropriate. There are, however, tracts of less hospitable and less developed lands which are not economically, at any rate, employed in agricultural production. These lands may be described as primarily the 'too much' lands—too wet, too hot, too dry, or too hilly. These lands lie at the lower end of the scale of man's settlement and developmental opportunities. Man, as would be expected, exploited as far as possible the more equable parts of the earth's surface.

As world population is increasing at an alarming rate, it is no surprise that man has had to search for increased production from existing developed land as well as attempt frontier development. In many parts of the world he has been able to exploit land—to escape from the bondage of the 'niggardliness of nature'. The development of new lands is closely tied to changes resulting from research as listed above.

Some of the few attempts made to develop new lands have been spectacular—if not in success, then at least in failure. Henry Ford's Fordlandia, a gigantic concept to develop Brazilian rubber, has already been mentioned in the section on plantation agriculture. It is appropriate here to mention that planning is sometimes conducted by private enterprise or in conjunction with governments. The Fordlandia project failed through lack of understanding of both environmental and social factors. Perhaps the pitfalls of the tropical world are of greater degree than those in temperate lands, and it can be expected that these lands will be developed last if at all. Yet there are conspicuous successes evident, even if they are not microclimatically tropical. Indonesian and Malayan plantation development in recent years has progressed into virgin rainforests, though the degree of scale is admittedly limited. In parts of Latin America, sheer pressure of migration has forced the exploitation of the Selvas.

Frontier development is by no means confined to attempts in the tropics. Perhaps the best example of non-tropical frontier development is seen in the non-chernozem lands of Siberia—the Virgin Lands Scheme for the opening up of vast expanses of eastern USSR. This scheme, implemented in the 1950's during Kruschev's regime, focused attention on new lands. Once more, as in the case of Fordlandia, it also focused attention on the difficulties inherent in such undertakings.

Another example of new lands development where failure is apparent is the Humpty Doo Rice Scheme in the Northern Territory of Australia. This scheme could be described as an abject failure after only a few years of exploitation, despite careful scientific and economic studies.

This is not to say, however, that all attempts at development are necessarily failures. The Brigalow Scheme in Queensland is a case whereby formerly indifferent land is gradually being transformed into an area which should become economically viable provided careful planning and research is used.

The question of new lands inevitably raises the concept of marginal lands. Experience in the development of virgin lands has brought home, often harshly, the dangers of farming to or beyond the environmental and ecological limits. Perhaps the earlier attempts have been too hasty; nevertheless, with the increasing pressure of population on land and uneconomic centralisation, many governments continue to look carefully, if not hopefully, at opening new lands. The progress of time should undoubtedly see more success.

ii. Rehabilitation of Adversely Affected Lands

Adversely affected lands are those lands which are subject to recurrent damaging and even catastrophic disturbances such as drought, flooding and erosion. These must be differentiated from areas affected by maladjusted land settlement patterns, where land has been abused by inappropriate exploitation, but obviously the two are often closely connected.

It is obvious that such lands occur commonly throughout the world with the resultant loss of much-needed production and economic hardship for many farmers. Australia provides many examples, being the driest continent yet with areas subject to recurring flooding. Much of its most productive agricultural land falls into this category. The Hunter Valley in New South Wales is a case in point. Although one of the most productive primary areas in the state, the Hunter Valley has been plagued by disastrous droughts and floods, one often following the other in rapid succession. Although the need for large-scale government action has long been evident, flood mitigation and drought prevention measures have only been piece-meal. This is perhaps a case where need has outstripped available funds.

Perhaps one of the best examples of government action in sustaining afflicted lands is found in the Netherlands, where large-scale projects to drain highly productive and otherwise unusable land are paramount. The traditional windmill, dyke and canal are ample evidence, as seen in the Dutch polders.

It is in these regions also that the gap between desirability and implementation is often great, as schemes are often very costly. Plans may be extended beyond the limits of agricultural development into the realms of hydro-electric power development, the re-siting of whole townships, the creation of transport networks and a re-orientation of other activities of a secondary or tertiary nature. It must also be noted that all too often governments act only after a major disaster or series of disasters; for example, drought leading to famine in Bihar State, India.

iii. Policies of Correcting Maladjusted Settlement Patterns

Throughout man's long history of settlement, he has displayed great ingenuity in using the land for his benefit. By the same token, however, he has been guilty of misusing and abusing the land through ignorance or over-zealous exploitation. Coupled with this is a dismal record in awareness and implementation of conservation. Some of the ramifications of maladjusted land utilisation have yet to be appreciated and it is only comparatively recently that man has recognised and attended to the question of wise utilisation and conservation. Few, if any, countries have escaped the ravages of man, particularly by early pioneers.

Misguided land utilisation was almost a concomitant of early frontier expansion, particularly in the New World countries. Overstocking and prolonged fallowing followed by natural disasters such as drought led to severe pasture deterioration in many parts of the world. Although short-term damage is easily assessed, the long-term effects are much more difficult to gauge and indeed may not even be recognised.

Many early forms of land use, such as lumbering, were conducted on a 'robber' basis and thus vast tracts of land were unnecessarily despoiled. It must be noted that the nature of land deterioration is such that regeneration may take decades, if it takes place at all.

Other forms of maladjusted land settlement include uneconomic farm sizes, exhaustion of soil nutrients, improper land tenure and the production of crops which cannot be integrated into the national economic structure.

It is not the place of this chapter to attempt detailed studies. However, briefly, three examples of government action in correcting maladjusted settlement can be seen in the Western Division of New South Wales, the dairy industry of Australia, and in the Tennessee Valley in the United States.

The Western Division of New South Wales comprises 41 per cent of the state and has a particularly high drought frequency; even in good times, the economic risk cannot be considered low. In the early phases of settlement, namely in the nineteenth century, the pastoralists expanded westwards, and in ignorance and greed, greatly overstocked the mediocre natural pastures with sheep. There was a dramatic increase in sheep numbers, climaxing about 1895. At that stage a cumulation of poor land tenure, overstocking and drought brought virtual disaster to the Western Division and the sheep population plummeted. The pastures had degenerated, soil had eroded, and settlers faced economic ruin. This led in 1900 to a Royal Commission, which produced the Western Lands Act of 1901. This act was further amended as some of the problems lingered. The essence of the Act was to rationalise the land use by way of closer settlement and pasture regeneration, thus ensuring a wiser response to the resource base. Even today, the setbacks of that period are still evident and government scrutiny and action are still common.

The dairy industry of Australia is clearly a case of land use being a response to social and economic forces rather than a prudent physical response. Very simply, there are too many uneconomic units in the industry and rationalisation is necessary. After many years of ineffective government subsidy which failed to attack the reasons for this inefficiency, the government has now decided on a scheme whereby it will buy out the economic unit at a fair price to the farmer and hence either lease it under new sub-division or channel it to other uses.

The effect of government action in alleviating maladjusted land settlement in the Tennessee Valley of the United States is very amply displayed. This will be dealt with further later on.

iv. Resettlement Opportunities

The creation of resettlement opportunities is closely allied with other government action, such as the development of new land, rehabilitation, projects such as irrigation and the application of new techniques.

A good example of this sort of scheme is the War Service Land Settlement Scheme, introduced by the Australian Government to settle discharged war veterans. This was usually associated with a developmental project such as the Murrumbidgee Irrigation Area. After some initial difficulties, this area can be considered to be an outstanding success as it accomplished the objectives of land improvement through irrigation, soldier re-settlement and closer settlement with higher productivity. Since 1945, 9,155 ex-servicemen have been settled on the land under the War Service Land Settlement Scheme.

Perhaps it is appropriate to note at this stage that not all government action is well-advised and successful. Some of the soldier settlement farms have come in for criticism. Chief criticisms are of the size of the farm and the provision of capital assistance, while there are indications that some individuals do not have the necessary qualifications. Perhaps this is true throughout all aspects of agriculture, as there are always people who fail to make good though their neighbours with similar holdings prosper.

v. Changes Resulting from Research

Changes in the land use patterns arising from the application of research successes provide some of the most spectacular developments in agriculture. There are numerous examples throughout the world, each case demonstrating the progressive ingenuity of man. With the dramatic increase in world population, there will be increasing pressure to develop means of more efficient agricultural production.

The most common developments are new or improved crop—livestock strains and types and the application of trace elements. Inalmost every case, research is conducted by governments either through their direct agencies or through the support of research and extension services.

An example of this can be seen in the Ninety Mile Desert in southern Australia. The government, through the Commonwealth Scientific and Industrial Research Organisation, discovered that this barren region could be made productive by the application of trace elements. This planning and investigation took many years, but out of it was born the Coonalpyn Downs Project which has transformed the wasteland into a rich pastoral area. This case is detailed in a further chapter of this book.

Research into new methods and forms of food production promises to change the entire land use pattern of the earth in time, and in this governments must bear an immense responsibility.

vi. Economic Sanctions and Restrictions

This form of government action is quite often indirect rather than direct in its effect on land use. Perhaps the most common occurrence is found in the international agreements which govern the trade of certain commodities. By this action, governments may protect, encourage or even discourage the choice of land use. There are, however, cases when the government has a direct and very active part in decisions concerning agriculture, as in the case of sugar farms in Australia. Here the government effectively controls output through the Sugar Board by assigning a quota of sugar to each mill which the mill in turn translates into cane quotas for growers.

Very often, government action in this regard is directed more at the function of agriculture within the economy rather than the direct interest of the farmer.

This latter point raises the question of the various levels at which government action may be taken.

2. Levels of Government Action

As mentioned earlier in this chapter, government action is manifest in many forms of agriculture. A corollary to this is that the levels at which governments may act also vary according to the stage of the economy and the type of government, as well as the function of agriculture within any economy.

Typically, in a free enterprise system such as Australia or the United Kingdom, the levels of government action are primarily at the State Government or County administration level, as the case may be, in co-operation or conjunction with the Federal or National Administration. Together with this are found semi-government bodies as well as co-operation with private interests. In a centrally planned economy such as in the USSR or Communist China, the initiative rests ultimately on the central government with local bodies merely carrying out instructions and seeking the objectives of the central government.

Irrespective of extent or type or level, the principles are broadly the same—to effect or superimpose an improved form of land use on the existing pattern of the land.

i. Government Planning on a National or Large Scale

National planning by governments is virtually synonomous with large scale planning, and is normally carried out on a national, federal or central government level.

The obvious examples to be considered here are the Socialist countries, where agriculture has become, or is becoming, the function of the state. The USSR has established a planning authority vested in the Department of Agriculture, which wields enormous power. Decisions such as what is to be grown and where, the methods to be employed, and the eventual marketing, depend on central government authorities. Planning is carried out to its most extreme degree and greatest scale in such a system.

Planned agriculture in the USSR dates from 8 November 1917, when the Second Congress of the Soviets decreed that private land ownership be abolished. Sale and purchase of land was prohibited and some 375 million acres of land were confiscated and passed into state ownership. The civil war of 1914–20 and the First World War did much to break down the existing pattern of agriculture. Up to 1927 the marketable value of the peasant farms was very low. Steps were then taken to reform this situation.

From 1927 on, the 24 million individual holdings of Tsarist Russia were gradually absorbed into either large collectives or state farms. The collective Kolkhozes today number in the region of 37,100, while the larger state farms called Sovkhozes number 12,200. Thus the first result of rationalisation was a spectacular rise in the size and organisation of agricultural units, and naturally in the work force per unit. The need for administration—planning the application of labour units to land units and its ramifications—was enormous. When the size of the units and the size of the work force are considered—the Kolkhozes average about 6,000 acres and employ about 400, while the Sovkhozes average 15,000 acres and 650 workers—then it is obvious that considerable degrees of planning are necessary.

It is difficult to evaluate the impact of planning in the Soviet system with any yardstick of private ownership. Certainly the figures from some Five Year Plans have at times revealed spectacular rises in production. On the other hand, there has been criticism levelled at the degree of planning, so that these figures are achieved at the expense of the administrative machinery involved in collectivism.

Certainly planning on a large scale is not confined to the centrally planned economies. There are good examples to be found in the free enterprise economies. A good example of such can be seen in the United States, whereby almost every facet of agriculture is under constant review by the Department of Agriculture, a department which wields enormous influence in the direction and function of American agriculture.

Such controls in a free enterprise system can take a number of forms, many of which are considered to be indirect means. These include price fixing at either end of the scale, guaranteed minimum prices or quotas to encourage or restrict production, subsidies offered as incentives, tariff protection against or even total prohibition of imports, levies for price stability and for the promotion of markets, government sponsored land settlement schemes, closer settlement, irrigation schemes and leasehold or financially assisted land development, and some measure of control through research and developmental facilities, free services and the like.

ii. Planning on a State and Local Scale

Planning on a state or local scale has had profound effects on the pattern of agricultural land use in nearly all parts of the earth. In many cases it is closely allied to federal planning. In the case of the centrally planned economies, such bodies usually carry out the policies of the central government as is found in the communes of Communist China. In the case of the free enterprise systems, such bodies often adopt an independent direction and the individual farmer is free to implement his own decision, although he will seek the assistance of such bodies as may exist.

The Department of Water Conservation in New South Wales is an example of a body engaged in planning at this level. This department controls and regulates the release and conservation of dam water to farms and thus has a powerful influence on the pattern of land use. Other state departments include Agriculture, Forestry and Soil Conservation. Other levels of planning include local bodies such as farmers' co-operatives or commodity boards such as the Milk Board.

There are also outstanding developments in agriculture which are the result of conscious planning by individuals—either in co-operation or acting individually. One such notable Australian example is presented by Yeomans' 'Keyline' concept of conservation and agricultural expansion based on an individual farm.

Perhaps the above points and their complexity of interrelationships can best be brought out by looking at the following comments issued by the Bureau of Agricultural Economics, Canberra (*Rural Industry in Australia*, p. 38/9):

Australia's major domestic policy objectives include the rapid development of resources as a means of sustaining a high rate of growth . . . and rising living standards . . . within the broad framework of its general economic objectives, the Commonwealth Government aims to promote a balanced development of agricultural resources, and measures directed towards these aims fall into two main groups:

(*i*) *Measures designed to increase production and productive efficiency.*

(*ii*) *Measures designed to stabilize returns for farm products, to market such products successfully, and to improve farm incomes.*

. . . *With the exception of wool, the marketing of all major rural products is subject to arrangements involving some degree of statutory control, particularly in the export field. Statutory bodies have been established at both Commonwealth and State levels.*

. . . *Commonwealth statutes provide for eleven commodity boards which are concerned with wool, wheat, dairy produce, eggs, meat, canned fruit, apples and pears, dried vine fruits, wine, honey and tobacco.*

. . . *Apart from the Commonwealth boards, there are over seventy marketing boards operating in the various states under State legislation.*

Government measures (as revealed by the Bureau) affecting rural industries in Australia include:

1. Research: this includes the services of the CSIRO and the Bureau of Agricultural Economics.
2. Extension Services: this involves the practical dissemination of research findings to the farmer.
3. Financial incentives: these include depreciation and investment allowances, deductions from income, zone allowances, differential taxation and customs tariff.
4. Rural Credit: these measures include preferential interest rates, Reserve Bank advances, term loan funds and Commonwealth Development Bank loans.
5. Public works: these undertakings provide the most conspicuous forms of government action in rural development. Examples include beef cattle roads, brigalow development in Queensland, the Ord River Project and the Coleambally Irrigation Area.
6. Land settlement: the War Service Land Settlement Scheme is a good example.
7. Special services and facilities: these include quarantine and pest eradication.
8. Fertiliser bounties: bounties are paid to stimulate the use of fertilisers to increase production. The Phosphate Fertilisers Bounty Act, 1963, is a good example.

Conclusion

Throughout most parts of the world, while farmers are generally free to decide on the nature and amount of their production, they are subject to government action because of the peculiar importance of agriculture. Thus the farmer must work in a broader framework than just a response to his physical and economic base. Social and political forces must be accommodated.

It has been shown that governments may intervene for a multitude of reasons and that implementation may take on many forms. It is difficult and sometimes dangerous to categorise these manifestations as the complexities of the various forces at work are more than evident.

Perhaps it would not be wrong to note that as population pressure throughout the world increases and governments become more and more centralised and take on increasing responsibilities in the distribution of resources, the impact of government planning is likely to increase rather than decrease.

Bibliography

Chisholm, M., *Rural Settlement and Land Use*, Hutchinson, London 1962.

Cohen, R. L., *The Economics of Agriculture*, Cambridge University Press, London 1957.

Murrumbidgee Irrigation—A Regional Study, *C.A.B.*, Vol. 31, No. 10, 1 April 1963.

Rutherford, Logan and Missen, *New Viewpoints in Economic Geography*, Martindale Press, Sydney 1966.

Schulz, T. W., *The Economic Organization of Agriculture*, McGraw-Hill, New York 1953.

Questions

1. 'Most agricultural areas of the earth show the geographical impact of planning.' What do you understand by this statement ? Illustrate your answer with examples.
2. What are the various objectives in planning ? Give an example to illustrate each type.
3. How may governments take the initiative in planning ?
4. To what extent does the stage of the economy and the type of government influence planning ?
5. Select a region where the impact of planning is conspicuous and carefully discuss the reasons, origin and measures adopted by the planning body or bodies.
6. How does agricultural planning in Australia differ from that in the U.S.S.R. and Communist China ?
7. 'Throughout most parts of the world, while farmers are generally free to decide on the nature and amount of their production, they are subject to government action because of the peculiar importance of agriculture.' Critically discuss.
8. Not only is there often a gap between desirability of action and implementation, but very often governments are hesitant to take any initiative. Why is this sometimes the case ? Give examples to illustrate your answer.

16 The Coonalpyn Downs Project

1. Introduction

The Australian Mutual Provident Society, Australia's largest and oldest life assurance organisation, commenced a project in 1949 on the Coonalpyn Downs, an area in the South Australia-Victoria border district previously known as the Ninety Mile Desert. This type of undertaking, which involved the development of virtually unproductive land, had not previously been attempted in Australia by a private financial institution. Because of the magnitude of the scheme, and its location within Victoria and South Australia, the governments of both states became involved—not directly with financial assistance, but with leasehold land grants and a great deal of help of a specialist nature. The federal government was also involved through the Commonwealth Scientific and Industrial Research Organisation's research activity and experimental data. The Coonalpyn Downs Project well illustrates the type of planning where private firms and governments combine their resources, though it must be stressed that the AMP Society was the main force behind all aspects the scheme initiated.

Basically, the scheme involved the taking up of large areas of almost totally unproductive land and converting it to the prosperous farms that exist today. The unproductive nature of most of the Ninety Mile Desert was not due to rainfall deficiency, but in great part to deficiencies in a combination of soil nutrients, particularly copper, zinc and phosphates. The minerals involved are termed 'trace' elements because of their presence in most soils in minute traces. It was obvious from research carried out by D. S. Riceman of the Commonwealth Scientific and Industrial Research Organisation, that a correction of these deficiencies could render the land highly productive. Most of this research by Riceman had been carried out between 1945 and 1948 and had been reported in journals and bulletins of the CSIRO.

The AMP Society considered that such a project would prove to be a satisfactory outlet for its invisible funds, and they were prepared to undertake the development of the land and the investment in plant and equipment at a time when suitable projects in other fields were relatively difficult to find. The Society has always identified itself with investment in primary industry, and took the decision when the Society's Board was informed of the investigations carried out by the Federal government body, the CSIRO.

2. The Physical Background

The Coonalpyn Downs extend from the Coorong in South Australia to the Victorian border in the east, while the southern limit is around Canniwigara. In Victoria a similar area—the Big Desert—extends eastwards from the border. Within this area the AMP was granted leasehold over 250,000 acres, with a proviso that additional

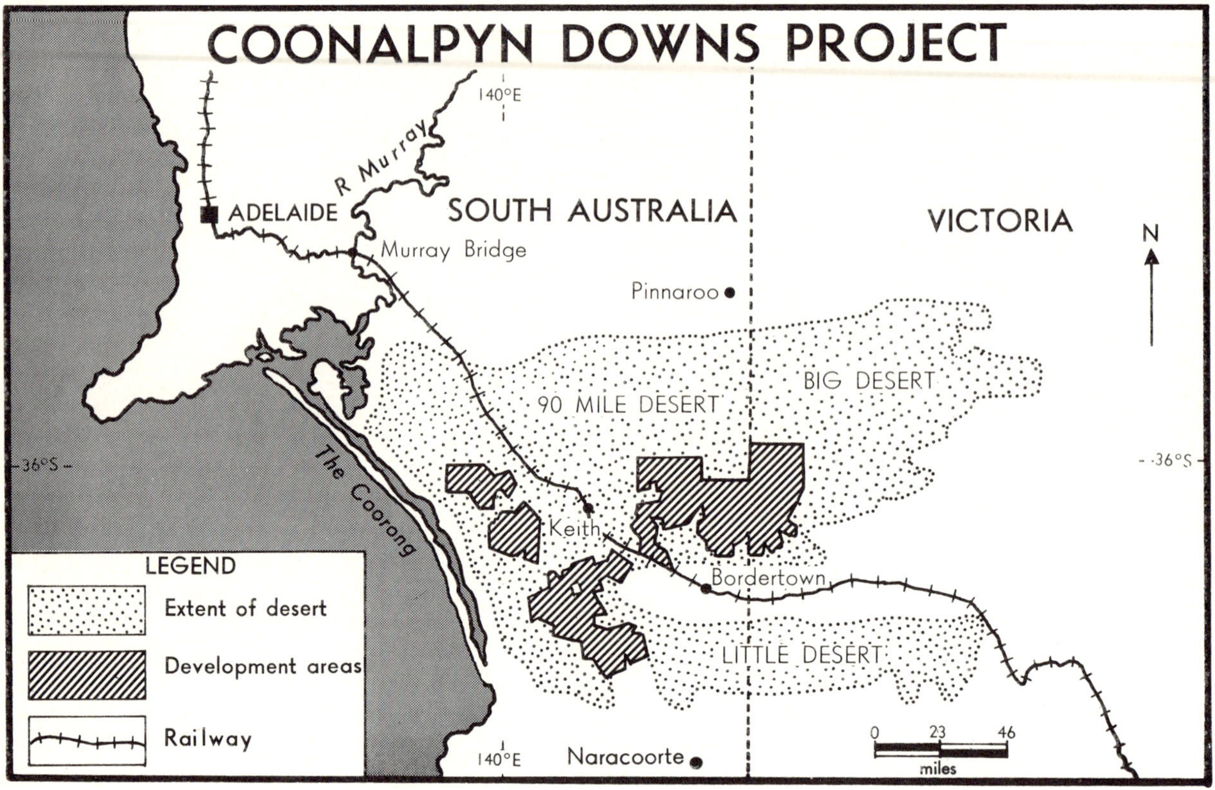

land could become available. This land grant was made possible when the South Australian and Victorian governments passed the necessary enabling Acts.

i. Topography and Soils

The topography throughout the region is gently undulating to flat, with a relatively low altitude of from 100–300 feet. Because of the gentle slopes and porous soils, surface streams of any permanency are unknown and only minor non-perennial watercourses are found. At close quarters the surface reveals one aspect which presented a problem, the uneven nature of depressions and rises of only a few inches height. These were to present difficulties particularly in the preparation of the seedbed, as is indicated later.

Soils are principally sands of varying depth over clay subsoils. The differences in the depth of the sand topsoil was of considerable importance in the success of development, since it was found that where the topsoil depth was around eighteen inches conditions were most suitable. Like many Australian agricultural soils, these were heavily leached, and commonly lacked phosporous, sulphur and calcium. These deficiencies are normally remedied by applications of superphosphate, as where similar conditions exist in most of the wheat lands lying close to the Ninety Mile Desert. The additional problem which gave the region its desert-like appearance was a deficiency in the trace minerals copper and zinc. Such a deficiency will not be remedied by the application of superphosphates.

ii. Natural vegetation

The original vegetation of the region was a low heath-like ground cover together with heavy concentrations of mallee (a dwarf eucalypt), other stunted eucalypts and a few species of xerophytes. The vegetation varied in height from about five to fifteen feet, and large trees were almost entirely absent. This made mechanical clearing relatively easy. Some variations of this soil-vegetation pattern occurred, and it was found that the land most suited to development was lightly timbered with associations of mallee and broombush (melaleuca).

iii. Climate

The region experiences cool to cold winters and hot summers. The winter temperatures resemble those of the surrounding wheatlands, averaging just below or above the 50°F. mark in June, July and August. Frosts are fairly predictable and less severe near the coastal margins, though even in the interior they are slight. Despite a lack of protection from winter winds the conditions can not be regarded as dangerous to stock if reasonable precautions are observed.

Selected Statistics (*temperatures in degrees F.; precipitation in inches*)

	J	F	M	A	M	J	J	A	S	O	N	D	Year
Keith, S.A. (40° 50′S; 140° 12′E.)													
Temp.	75	75	72	65	59	54	50	52	58	61	68	72	66
Precip.	0·7	0·9	0·8	1·3	2·1	2·0	2·1	2·1	2·1	1·6	1·3	1·0	18·0
Bordertown, S.A. (41° 05′S; 140° 30′E.)													
Temp.	72	73	71	66	58	55	52	54	59	60	60	67	62
Precip.	0·8	0·9	0·8	1·5	2·0	2·4	2·3	2·2	2·1	1·8	1·3	1·1	19·2

Scientist David Riceman of the CSIRO who led the team that discovered the trace-element deficiency on the 90-mile desert takes a segment of sample plot of clover for laboratory tests. A rectangular wooden frame (shape visible behind the paper bag) is pressed down into the clover and the section is cut out to give exact comparative growth samples for the different plots.

Typical undeveloped scrubland on the 90-mile desert plain. The foreground shows the type of low sand dune formation which required different treatment from the level plainland. The vegetation is mixed scrub with scattered clumps of whip mallee.

Summer temperatures can be hot except when moderated by coastal influences. They average in the mid seventies in January, February and December, and many days are close to the century in late January and February. The season of high evaporation occurs during the rainfall-deficient months, and high evapotranspiration rates can adversely affect pastures. Water conservation is also important.

Precipitation over the whole area varies from around 16–20 inches per annum, with a winter maximum accounting for over two-thirds of the total. Thus the rainfall occurs mainly during the most effective part of the year for maximum plant growth, and resembles the Mediterranean environment in this respect. No months are without rain, although January, February and March have means less than one inch. The figures for both Keith and Bordertown (above) emphasise that the 'desert' nature of the landscape was a response to the low fertility status of the soils rather than a paucity of rainfall.

3. The Plan

In 1949 the AMP Society took over an area of land that was largely undeveloped. Two small bases were available, consisting of several hundred acres and a house, but other capital developments were very few; roads and fences were practically unknown and little surveying had been carried out. The Society was handicapped by the fact that no precedent for developing such an area existed. The War Service

Land Settlement Scheme, which had involved extensive areas in successful projects, was based on partly developed country within the range of 'normal' agricultural conditions, where soil types were relatively much more fertile.

The first stage was an appraisal of the land resource. Surveying was carried out by aerial methods and in the field, and ecological and soil plans were developed. When the plans had been assembled divisions were made of the first 30,000 acres, this being the amount the Society hoped to develop annually. The acreage was to be handled in 10,000 acre lots from each of three bases, each base being under the control of an Area Manager. The total area programmed for any base was 50,000 acres in a five year period. This would mean that each base would be developing—in various stages—part of the 50,000 acres, while these preliminary tasks were being carried out, men were being recruited to work on the several aspects of the scheme. These people were to become the landholders on the developed properties as they came to completion. At the same time houses and sheds were constructed for the staff and barrack type accommodation and facilities for some of the men. Agricultural implements which were best suited to the conditions of the area were obtained and the necessary facilities to put the scheme into operation were finalised.

4. The Methods

The consideration always before the planners was the evolving of techniques to suit development on a very large scale. Whereas individual plans are usually carried out in conjunction with a developed area (that is, the rest of the farm), methods in the Coonalpyn Downs were directed towards the strictest economies, since no income was being earned and no tax remission benefits such as are available to private landholders were granted. The need in all aspects of development was to employ the cheapest and quickest methods possible, without loss of efficiency.

i. Clearing

The first job was to knock down the scrub. In the beginning logs and railway lines hauled by 40 horsepower tractors were used, then gradually wider logs and tandem arrangements of several logs were employed. Finally some three hundred yards of ship's anchor chain hauled between two 140 h.p. tractors was knocking down approximately one hundred acres of scrub per hour. The cost of 'scrub bashing' with these methods was much lower than could be achieved using smaller conventional implements.

The levelled scrub was then allowed to dry out over summer in preparation for burning during the autumn. Every effort was made to achieve a clean burn to give the best possible surface for ploughing. Fire breaks were carefully prepared and the land to be fired was divided into areas to be scrutinised by the strategically located fire control units. On the day selected for burning a lighted wick was drawn behind a Land Rover vehicle and rapid lighting was effected.

ii. Seedbed Preparation

Although in the majority of cases a relatively clean burn resulted, beneath the surface thousands of mallee roots and stumps remained. These would quickly 'sucker' or produce shoots at ground level, as the roots of this eucalypt commonly do unless they are extracted. So the land had to be ploughed to a depth of about nine inches. Again heavy machinery was found to be quick and relatively cheap, and

Majestic Ploughs, which are equipped with eight heavy duty discs of thirty inch diameter, were pulled by crawler tractors and successfully ploughed out the bulk of the mallee roots.

Harrowing the ploughed surface to produce a seedbed also presented difficulties. It was found that conventional stump-jump harrows were not fully effective to overcome the rough nature of the country and the thousands of mallee roots which now lay on the surface, so a frame of welded railway irons was fabricated. These were hauled over the ground, levelling and consolidating it and dragging a few more mallee roots to the surface for burning.

iii. Seeding and Topdressing

Sowing the pasture seed was also accomplished by use of large scale methods, since the uneven nature of the land and the need to sow some thirty thousand acres per year rendered conventional seed drills impracticable. Conventional broadcasting units were mounted on pairs of frames allowing seed and superphosphate to be spread over a thirty-six foot swathe. Each unit could thus sow and fertilise between one and two hundred acres per day.

Where frequent variations in soil types occurred it was necessary to use a unit which carried two seed boxes, each box holding a different mixture suitable for either the flatter areas with shallow topsoil or for the sandy rises. The tractor driver could select the operation of either box from the tractor seat without halting. Seed covering was usually carried out at the time of sowing, and again anchor chains were frequently used to cover the seed and compact the soil.

Rates of seeding and application of fertilisers varied according to soil types and topography. In general the rate of application of superphosphate was three bags per acre at sowing, and a further topdressing of one bag per acre after pasture establishment. The total application of high grade superphosphate was thus one third of a ton per acre (1 bag = 187 pounds). The vital trace elements were added in the form of copper sulphate and zinc sulphate at the time of sowing and they totalled seven pounds per acre.

Typical seeding rates were at four and three quarter pounds per acre, made up of four pounds of legume seed and three quarters of a pound of grass seed. The legumes included mixtures of subterranean clover (commonly Mount Barker, Bacchus Marsh, Clare, Yarloop and Dwalganup varieties), strawberry clover and lucerne. The chief grasses used included Phalaris, Perennial Rye Grass and Veldt Grass. Mixtures of these were varied according to soil conditions.

iv. Farm division and allotment

During the period after seeding divisions were made into 'living areas'. The living areas naturally varied according to the nature of the land, but the plan was to create areas which would carry the equivalent of 1,500 dry sheep when fully developed. To allow for some expansion, scrub land was usually included on each block. Basically each farm on completion had four or five paddocks, a minimum of three fully equipped watering points, a house and a shed. As applicants selected their future properties they were allowed to express opinions regarding subdivision, location of water points and style and size of the house. Where possible the Society reflected such wishes in their subsequent actions. The applicant had previously been given a wide choice as to the location of his property.

As in all aspects of the project, methods of division were aimed at speed and maximum efficiency. The Society operated its own water boring plant, and machines for post-hole digging and boring, for trenching and other jobs were widely used.

Allotment of farms was based on seniority, that is the date of enlistment of the men as workers into the scheme. Groups of properties which became available at the same time were offered to the senior group of men under a ballot scheme. Each property was valued as it was completed and offered to the applicant, who had absolute freedom to accept or reject. When, as was invariably the case, he accepted the block, he was given financial assistance from the Society to purchase stock, implements and other necessities.

5. The Impact of the Project

The AMP Society warns that there can be dangers in evaluating land development schemes, due mainly to exaggeration of the size of the project, the capital involved and premature anticipations of success. The Coonalpyn Downs Project has been subject to some extravagant reporting, particularly in the early period. Despite this, there are substantially based indications of the success of the project in both terms of agricultural output achieved and in the stimuli on other sectors of the community.

'Before and After' on the 90-mile desert. To the left of the netting fence is the worthless scrubland, while on the right side is the vigorous pasture established on trace-mineral enriched soil. The level nature of the land is obvious, with the low residual hills visible on the horizon.

J. E. Coaldrake reported in 1951 in a CSIRO *Bulletin* that 'except for pockets of better country, such as those at Keith and Coonalpyn, the Ninety-Mile Plain supported only a sparse population until quite recently. The chief industry was the raising of sheep, which were depastured primarily on the ephemeral herbage that follows burning of the scrub. On this basis, the country carried approximately one sheep to forty acres. This practice was supplemented with limited areas of sown pasture, in which *Oenothera odorata* (evening primrose) was the principal species. Cereal cropping was attempted but it was unusual to harvest more than two crops before the soil was exhausted. In places, even the first attempt to crop failed.'

From land such as this the AMP was able to report in 1964 that some 425,000 acres had been converted into 200 first class farms. The established pasture is capable of carrying in excess of one breeding ewe per acre. On the majority of farms a limited amount of cropping is carried out, though the aim is usually to assist in pasture renovation and allow fodder conservation. Some dry sheep and beef cattle are also grazed, the percentage of cattle to sheep being usually restricted to 5 per cent, however. The farms have achieved in average seasons a very satisfactory gross income, which has allowed a considerable margin for meeting debts and allowing financial progression. Wool cuts have been consistently high, lambing percentages are good and fat stock well regarded at local markets and in Melbourne, Adelaide and Ballarat.

Although these achievements are agriculturally spectacular, of great interest also is the impact on other aspects of the Coonalpyn Downs area. The rural towns of Keith and Bordertown have expanded since 1947 when the Society began preliminary work—Keith, for example, showed a population growth from 519 persons in the 1947 census to 1,126 in the 1954 census, while the old-established centre of Bordertown progressed rather slowly from 1,396 to 1,847 persons in the same period. The state electricity grid was extended to both towns, a reticulated water scheme from the Murray River to Keith was engineered and this town built a post office, police station and hospital. Six stock firm offices and seven banks were opened in the two towns in the period 1946–55. The numbers of school children attending Keith Area School rose from 64 to 1947 to 338 a decade later, and in Bordertown the increase was from 201 to 564 in the same period. Most of the developed area of the Project lies within the County of Buckingham, and official figures for the period 1954–55 and 1956–57 for sheep and wool give clear indications of the continuing growth. In this period sheep numbers rose from 378,000 to over half a million, and wool production jumped from less than £4¼ million to just over £6 million.

The vital importance of superphosphate to the continuing fertility level of the soils is reflected in the tonnages handled at Keith and Bordertown railway stations. In Bordertown the amount rose from 3,495 tons in 1945–46 to 12,297 tons in 1956–57. In the same period the figures at Keith leapt from 1,264 to 20,837 to give the station the distinction of handling more superphosphate per year than any other country railway station in South Australia.

Harold F. Bell and W. H. Cairns of the AMP Society summarised the Company viewpoint of the 'National Benefits' of the scheme in notes originally prepared for the 1957 Wool Seminar at the Australian National University, Canberra, ACT:

It is the contribution which the AMP *Society's scheme is directly making to the nation's production of wool, and other rural commodities which is, at first glance, of most significance. As time goes by, however, it may well come to be recognised that, from a national viewpoint, the indirect benefits arising from the scheme are even more significant.*

In particular, there is the fact that, stimulated by the AMP's *initiative, and encouraged by what has been accomplished in a relatively short time, many private individuals (who are fortunate enough to have the further inducement of taxation concessions in respect of land development) are taking up land in the area. Some details were given earlier of the tremendous growth of business and other activity in the Keith and Bordertown districts. It would be absurd to suggest that this advance is solely due to the Society's project, but the latter must certainly have had a profound influence in this direction.*

And the influence of this particular undertaking has probably gone much further afield than this. The fact that throughout our history this area of reliable rainfall and excellent access to markets has hitherto gone undeveloped has undoubtedly inspired many interested parties to inquire if there are not other neglected areas in Australia, which, with the adoption of modern techniques, could be brought into production on economic terms. (Obviously the factor most likely to provide a check to such development waves is the decline in the level of prices for wool and other rural commodities such as is now being experienced.)

The sphere of indirect influence can be stretched even wider; for the scheme has attracted the keen and approving interest of important overseas visitors and has probably made a significant contribution to the creation of the climate of a "growth economy", which atmosphere is conducive to the attraction of overseas capital. For example, at the instigation of the Commonwealth Government, the AMP *project has been viewed by officials of the International Bank during their visits here.*

Land development is a risky business and the land developer must be prepared for the unexpected, sometimes pleasant, sometimes not. The AMP's *experience has been no different from this; some activities have succeeded beyond original expectations; in other phases there have been disappointments. But at this date the* AMP's *project appears to be the only instance in the history of this country where suitably equipped young Australians have been settled, and financed, on to good class properties of economic size, without involving huge governmental outlays.*

Bibliography

AMP Society, Leaflet 0025: 'The AMP Land Development Scheme', Leaflet 0026: 'Desert Conquest'.

Bell, F. B., and Cairns, W. H., 'Notes on the AMP Society's Land Development Scheme in South Australia', *Australian Journal of Agricultural Economics*, Vol. 2, No. 2, December 1958.

Coaldrake, J. E., *The Climate, Geology, Soils and Plant Ecology of Portion of the County of Buckingham (Ninety-Mile Plain), South Australia*, CSIRO Bulletin No. 266, 1951.

Commonwealth Bank of Australia: *Building Soil Fertility in South Australia.*

Riceman, D. S., *Effect of Zinc, Copper, and Phosphate on Subterranean Clover and Lucerne grown on Laffer Sand near Keith*, CSIRO Bulletin No. 234, 1948.

Riceman, D. S., *Effect of Cover Crops and Phosphate on Subterranean Clover, Lucerne and Phalaris Sown with Zinc and Copper on Laffer Sand, near Keith*, CSIRO Bulletin No. 249, 1949.

Riceman, D. S., *Preliminary Investigations on the Laffer Sand, near Keith*, Journal of CSIRO, Vol. 18, 1945.

Rural Bank of NSW Journal, *Trends*, March 1957.

Questions

1. What were the main reasons for the AMP Society initiating proceedings which resulted in the development in the Coonalpyn Downs?
2. Give a brief outline of the environment of the Ninety Mile Desert area before development. What specific problems for agriculture did the region present?
3. In what ways did the scheme differ from other land development schemes?
4. Outline the methods adopted in the Coonalpyn Downs development scheme. In what ways did ingenuity assist the developers?
5. Explain fully the importance of artificial fertiliser and other additions used in this scheme.
6. Explain the procedures by which individual farmers obtained and developed their individual holdings.
7. Discuss fully the impact of the plan as a whole.
8. 'The Coonalpyn Downs Project has been vastly over-rated and at this time (1968) it is doubtful if the investment of great amounts of finance has been warranted.' Does the evidence available support this statement? Discuss the statement fully.

17 The Tennessee Valley Authority

1. Introduction

The Tennessee Valley Authority's scheme for the reclamation of the Tennessee River valley directly affects a considerable portion of the south-eastern United States. Areas within the states of Kentucky, Tennessee, Mississippi, Alabama, Georgia, Virginia and North Carolina are concerned. Here are the lands drained by the Tennessee and Cumberland rivers and their tributary streams, including the Clinch, Holston, Little Tennessee and Hiwassee.

The Tennessee Valley scheme resulted from a plan ratified by the national government and carried out by an Authority charged with its implementation. Thus the scheme was planned by a national agency, although much of the physical execution of the plan was in the hands of private enterprise firms. In some instances the scope of the scheme demanded special co-operation between government and private concerns as they tackled tasks of gigantic proportions. The scheme was not only immense, but far reaching in a variety of ways. It was primarily directed at only one aspect of the rehabilitation of the valley—the improvement of navigation facilities. The other major improvements of flood control, agricultural development, rehabilitation of forest lands, the development and distribution of hydro-electric power and establishment of artificial fertiliser plants grew out of the original scope of responsibility assumed by the Authority.

2. The Establishment of the TVA

i. Early Exploitation of the Valley Lands

The early settlers had cleared slopes in the valleys of the Tennessee's tributary streams and in the main valley, with little regard for the detrimental results. The widespread clearing greatly increased an already high percentage of runoff, resulting from a rainfall which everywhere in the catchment area is at least 40 inches and usually 50 inches or more per annum. Nashville on the Cumberland river, with 47·4 inches per annum, and Knoxville on the upper Tennessee, with 47·6 inches of rainfall annually, are representative stations of the area. Since the distribution pattern is of a winter maximum, or one of as much winter as summer rainfall, the runoff is greatest when soils are saturated in the cool, low evaporation period from November to March/April. Greatly increased runoff from the denuded slopes was the main factor causing the catastrophic floods which periodically swept the Cumberland-Tennessee river systems, as frequently as every year. The alternating flood levels and periods of low water made attempts at irrigation difficult, and in association with this erratic flow there was a continual increase in stream siltation, threatening to render river traffic impossible in places.

In the lower valley slopes the over-exploitation of farmland was as great as on the higher slopes; although here erosion was less of a problem, deterioration in soil

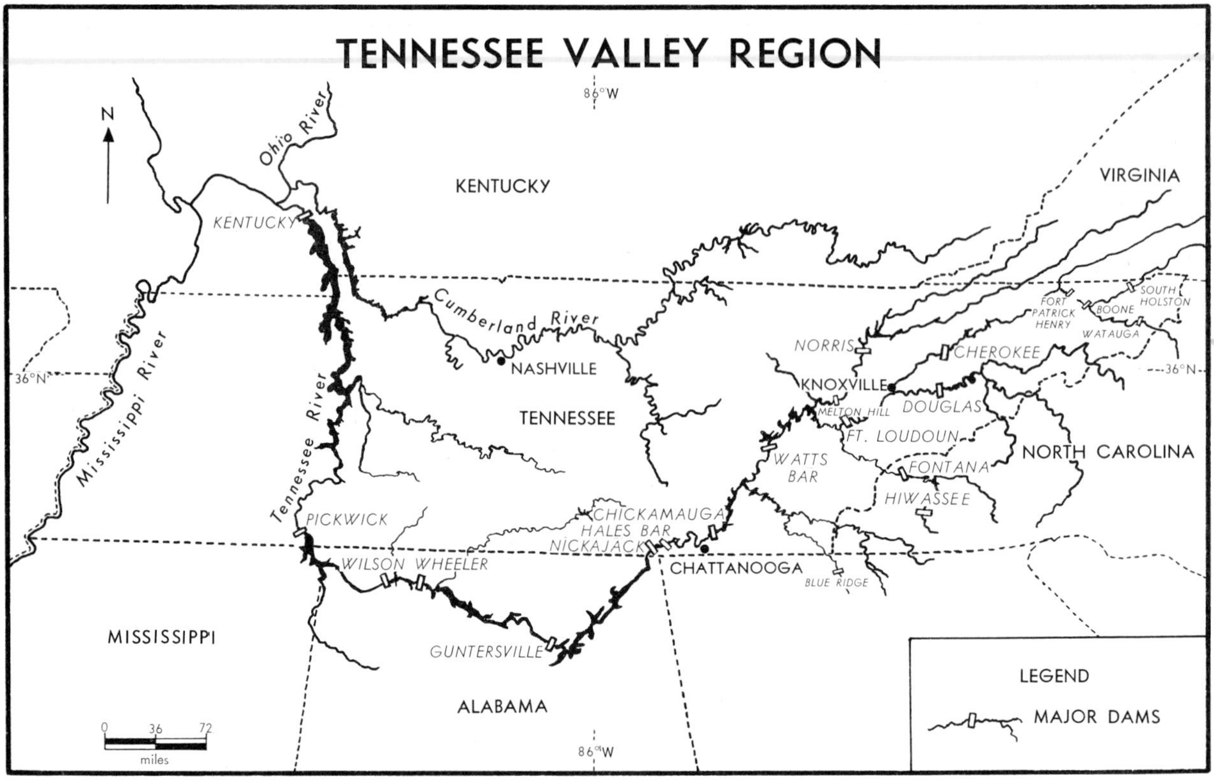

fertility was greater. The continued monocultural cropping of cotton, at first on plantations and later on tenant and sharefarmed estates, had left the land 'crop sick' and the soil depleted of nutrients. The topsoil structure was damaged by repeated ploughing and harrowing with a resultant loss of organic matter. The continuing decrease in yield of the land forced cotton to take over the rich silts along the river banks, creating unprotected fields and thus adding to the disaster of flooding when natural levees were breached.

The combination of soil erosion, flooding, siltation and decreased soil fertility meant that in 1933 the per capita income of the valley people averaged only US$168—less than half the national average. Only 3 per cent of the farms were connected with electric power; 9 out of 10 farmers could not afford to use artificial fertilisers. In any case many farms were too small for efficiency, averaging only 70 acres, an area which restricted rotation and mixed livestock and cropping activities. The great natural transport routes along the waterways were barely employed, only 33 million ton miles of freight being recorded on the Tennessee in 1933—a figure less than $\frac{1}{50}$ of the 1966 freight load, and possibly less than one per cent of the potential capacity.

ii. Early Moves towards Conservation

The 40,000 square miles which comprise the drainage basin of the Tennessee river presented a startling example of misuse of natural resources when the administration of President Theodore Roosevelt first attempted to reverse the strongly running tide of wastefulness. During his presidency from 1901–09, and due in large part to the advice and influence of US Chief Forester Gifford Pinchot and of leading geologist and hydrologist W. J. McGee, Roosevelt supported measures which were to lead to the first moves towards conservation.

The first Reclamation Act of 1902 was directed at restricting the private development of hydro-electric facilities along the nation's waterways. The defeat of a bill by veto in 1903, and subsequent amendments written in to the Reclamation Acts in 1906, further strengthened the Federal Government's control over power sites and provisions to sell power for municipal purposes.

In these activities Theodore Roosevelt played a leading part. He had vetoed the 1903 propositions which would have opened the way for private use of the hydro sites, and he strongly opposed the private development of the Muscle Shoals, Alabama, power station, which came to be regarded as the test case for the Reclamation Acts. Apart from these blocking moves, Roosevelt made one vital step towards the establishment of the TVA by creating the National Conservation Commission. The Commission was empowered to produce a comprehensive report on all aspects of the Tennessee Valley, a report which revealed a considerable amount of serious soil erosion, river siltation and depletion of ground cover, and created a basis of concern for the social welfare levels of the valley peoples.

The Commission was responsible for presenting a new viewpoint on the conservation of a region's resources. Most previous studies had been in the hands of organisations whose main interest lay in the political units—the states. The Conservation Commission viewed the Tennessee Valley as an entity, a geographic unit which overlapped the boundaries of seven states. Because of this the Commission presented a report which viewed the river valley as forming part of a unified development project embracing all aspects of the valley and its resources. It was this investigation which was to create the serious political issues which TVA raised in the 'thirties. The United States Constitution places strict limits on the Federal Government's powers of planning by refusing it a charter to act in the states' economies. Since the federal body is empowered to act in removing barriers to commerce between the states, by arguing that poor navigation facilities in a river system are a barrier, it may implement a project which will require a number of dams, locks, weirs and channels. The further requirements of the control of runoff, protection of the catchment area, elimination of unnecessary silting and other conservation aspects are not permissible; and the right to sell electricity from a project, for example, may also be unconstitutional.

The unified development concept was based on a multi-purpose scheme. The influence of Forester Pinchot was obvious, for he claimed that the major factor in restricting river flooding was the establishment of forests, and in turn lessening of erosion and provision of navigational facilities depended indirectly on forest establishment. Although the report was a major step, implementation of a unified plan was still a long way off. The leading figure who emerged in the 1920's was ardent conservationist Senator George Norris of Nebraska. Norris became the leader in saving the Muscle Shoals fertilizer production facilities for public use. Two Norris bills which would have opened the way for public operation of the plant were defeated by the Coolidge and Hoover administrations in 1928 and 1930 respectively. The concept of the unified development of the Tennessee was held back until the third Norris bill was passed by Congress on the 18th of May, 1933, during the administration of President Franklin D. Roosevelt.

iii. Tennessee Valley Authority Legislation

The 1933 establishment of the Tennessee Valley Authority was a vital plank in the Democrat platform, and an integral part of Franklin Roosevelt's 'New Deal'. As a political decision it has had many critics, who have based their dissatisfaction with the scheme either on the cost to some sectors of the community or actions of the Federal Government engaging in electricity and fertiliser sales, claiming these to be unconstitutional. In evaluating the success of the TVA as a piece of geographical planning, the political considerations must be left aside. To remain within the Constitution, Roosevelt used, as any Democrat or Republican president must use, a roundabout approach to the multiple problems of the Tennessee Valley. The basis of the Authority rested on the single plan to remove barriers to interstate commerce, by controlling the flow and navigational capacities of the Tennessee River. The considerable impact of the TVA also rests with the successes of allied aspects of its planning which resulted from this single constitutional activity.

In submitting the legislation to Congress in 1933, Franklin Roosevelt clearly pointed out that the TVA was 'charged with the broadest duty of planning for the proper use, conservation and development of the natural resources of the Tennessee River drainage basin and its adjoining territory for the general social and economic welfare of the Nation.' Thus was reflected the unified concept of thirty years before.

The Congressional Act laid down specific objectives:

Dams were to be constructed so that a navigational channel of sufficient depth (nine feet) was to be created in the Tennessee River to allow the passage of barge traffic, to control potential floodwaters in the Tennessee and lower Mississippi rivers and to regulate the seasonal flow of these waters.

A further objective was related to the provision of hydro-electric power plants for generation of domestic, agricultural and industrial power supplies and the further commissioning of chemical and fertiliser plants for the production of improved agricultural fertilisers and provisions of munitions manufacture in times of emergency.

Roosevelt, however, regarded the TVA as more than a material welfare and national defence programmed scheme. It was a vital part of his 'New Deal', and as such it incorporated idealistic social considerations that rehabilitation of such an area could bring forth. Thus, coupled with the enormous material changes the scheme has produced, there has been a great rise in social welfare values too. Despite the considerable pockets of poverty which still exist in parts of northern Tennessee and the borders of Kentucky, North Carolina and Virginia, standards of living throughout the Valley have risen dramatically. Taking the per capita income as a yardstick, the overall increase in the United States from 1933 to 1966 was 441 per cent; in the same period that of the population encompassed by the Tennessee Valley Authority rose by 624 per cent. Such results mean that the scheme has always had a tremendous appeal for developing economies in parts of the world where similar measures can be undertaken. Hundreds of foreign experts have organised research visits through the TVA headquarters at Knoxville, Tennessee, and TVA technicans and specialists have visited developing nations to advise on aspects of development in comparable schemes. India's Damodar Valley scheme, for example, has close ties with the Tennessee Valley scheme.

3. The Implementation of the TVA Plans

i. Erosion Controls and Valley Slopes

Control of the Tennessee River rested on two factors—the need to control the flow by a series of dams in both the main stream and its tributaries, and the check of headwater tract erosion and excess runoff.

Control of the headwater tract erosion has been a detailed effort resting on such

These two TVA photographs of the same area show (top) badly eroded farmland, and (bottom) a graded and grass grown streamline engineered to prevent further erosion. The deep gully in the upper picture not only took a large area of pasture from the livestock feeding area, but threatened injury or death to the animals. Erosion was rapidly spreading and threatened to engulf more and more useful soil. The same land two years later had been sloped and sodded at the direction of agricultural engineers. Terracing, sloping and a grass stream bed slows and retains runoff, preventing uncontrolled freshets from gouging out the soil.

physical checks as contour ploughing, strip cropping, grading of slopes, reafforestation of slopes, check dams, silt ponds and grading for example. The Authority was joined by two government bodies in arresting the erosion of the catchment area —the Soil Conservation Service and the United States Geological Survey. As a result of their activities, a thorough soil survey of the Valley's soils has been carried out as well as a complete topographic mapping project.

Tree planting was a major aspect of the work. Seedlings provided by the TVA nurseries were planted in thousands on the Valley slopes, so that today forest covers 54 per cent of the area within the TVA, yielding an annual income of US$450 million. A variety of trees, including Black Walnut, Black Locust, Honey Locust, Hickory, Cypress, Oak and the very important Southern Pine are valuable for such activities as the building industry, furniture manufacture and paper pulp making.

As Authority action was being carried out to arrest major soil erosion problems, the individual farmer was introduced to a range of services designed to assist in the combat of excessive runoff, and at the same time improve agricultural techniques. Fertilisers were available in increasing quantities from 1935 onwards as the Muscle Shoals plant came into operation. Various methods of strip cropping were demonstrated, as well as contour ploughing and banking and the provision of brushwood barriers to protect slopes; livestock numbers increased as pastures became established. Cover crops like the oriental legume lespedeza and a wide range of other plants were demonstrated and suggested as being suitable, and chemical control of weeds and pests demonstrated. With changes in the ratios of animals and plants and an ensuing diversification came prosperity levels which have enabled farmers to put back some capital towards maintaining permanent erosion controls.

ii. Flood Control and Navigation Facilities

In the main streams the river control has been the most spectacular achievement of TVA. The Tennessee is dammed at nine places in 650 miles, forming a series of gigantic lakes which permit barges to travel via the nine foot deep channel that has been created from Paducah to Knoxville. The system employs locks on the major dams, since the fall throughout the Tennessee is around 800 feet.

The dams are spaced in the following way throughout the valley (measured, miles above the confluence of the Tennessee River with the Ohio River): Kentucky 23; Pickwick 207; Wilson 259; Wheeler 275; Guntersville 349; Hales Bar 431; Chickamauga 471; Watts Bar 530; Fort Loudon 602. Knoxville is a further 48 miles upstream from the face of the Fort Loudon Dam wall, and is the limit for barge traffic on this section of the river.

In addition to these nine major dams, there are another twenty-two control dams in full operation, and three others, Nickajack, Tim's Ford and Tellico, are under construction. Five major dams are located on the important tributaries of the Tennessee—Norris on the Clinch river, Cherokee on the Little Holston river, Douglas on the French Broad river, Fontana on the Little Tennessee river and Hiwassee on the Hiwassee river. The Little Tennessee river is the site of the Alcoa (Aluminium Company of America) dams—the Chilhowee, Calderwood, Cheoah and Thorpe dams—and two others are found on the Little Tennessee tributaries, the Santeetlah and Nantahala dams. The construction of the Tennessee River dams was carried out with spectacular speed after Congress ratified the TVA in mid-1933. Norris and Wheeler dams were under way within six months, Pickwick and Guntersville followed in 1935; and as the first two were being completed in 1936, Chickamauga and Hiwassee were begun.

The major dams are multiple purpose constructions, aimed at controlling floodwaters, raising and lowering the level in the artificial lakes system for barge traffic, generating hydro-electricity and supplying irrigation water and supplies for domestic purposes and a multitude of industrial demands. To this end the system operates in response to runoff. At the beginning of winter, the level of the dams is progressively lowered to enable the additional floodwaters to be accommodated. Before the effects of heavy rains, water levels are also lowered—via the power turbines wherever possible, but through spillways and sluices if the overflow is too great for the turbines to handle.

The main flood season finishes in early April and after this the levels of the lakes are allowed to rise again. In the spring the rains fill the dams and reservoirs to their greatest height, in summer and autumn the water is released to produce power and keep the water level high enough for barge traffic. Releases of water continue until the next winter when the levels are low again, ready once more to cope with the ensuing high levels of runoff. The integration of the dams with one another in a controlled scheme is only part of the pattern, for the Tennessee system must also hold water back when it is necessary to restrict flow in the Mississippi river. Tennessee dams have been instrumental in keeping Mississippi flood levels down on several occasions, often a few inches below crucial heights.

The great inland waterway of the Tennessee is a result of both the success in controlling the headwater streams as well as the main river. The importance of river traffic can be gauged from the fact that a single tow-line of barges will shift the equivalent of 400 rail freight waggons. The importance is heightened because of the bulky nature of the materials being moved, including coal, oil, grain, chemicals, timber and various ores.

Prior to the establishment of the network of dams, there were impassable obstacles limiting traffic to certain sections of the river. Muscle Shoals, for example, was a treacherous section of rapids in low water and a raging torrent at high water levels. The Wilson Dam has now been sited on the shoals, thus permitting traffic to pass up river via a lock. The lowest dam, Kentucky, permits barges to enter the upper valley by way of its large lock lifting traffic 75 feet vertically.

iii. Power facilities and Industrial Production

Possibly the production of electric power has done most to lift standards of living in the valley, although at the same time it is the provision of this utility which has aroused considerable criticism of the Authority. The criticism is directed at the cost structure created by the TVA in selling power, it being claimed that the Authority is selling below cost of production. This type of argument is a perennial one in multi-purpose schemes where the capital costs of the project are spread over a number of consumers. In this type of projects the consumers may include river transport firms, riverside industries and industries drawing power from hydro sites, domestic power users, irrigation water users and all sectors afforded protection against flood. Power output from the Tennessee reached 12 million kilowatts in 1963, and is estimated to reach 19 million kilowatts in 1969. Practically the whole of the valley is now served with electric power. There are some 3,700,000 consumers here, as well as the peoples of most of Tennessee, large parts of Alabama, Kentucky and Mississippi and smaller parts of Georgia, North Carolina and Virginia. Power is distributed to an area of 80,000 square miles and to 5,300,000 consumers.

So great has the demand for power become, that thermal power stations have had

Typical Tennessee farmland showing evidence of sound conservation measures. In the background the restriction of grazing has maintained forest cover on erosion-prone slopes and is actively encouraging natural regeneration. The foreground steep slope is well protected by a grass ground cover. The large dams serve to supply stock and domestic water, and limited amounts of irrigation water. The small ponds are a source of income too, used to raise minnows which are sold as food fish for larger game fish. Finally, forest and ponds encourage wildlife, and this is a feature of TVA conservation planning.

to be built by the TVA, and they now account for three-quarters of the total supply of electric power. The power is employed in a wide range of industries, of which the following are some of the more than 1,600 which have located in the Tennessee valley since 1929. At Muscle Shoals there is the original power plant which has been developed, and this area also contains two major nitrate plants largely directed at supplying high explosives material; several types of armaments are manufactured and vast quantities of phosphatic fertilisers from phosphate deposits in the southern Nashville basin. Sulphuric acid for Muscle Shoals comes from the Ducktown copper smelter which receives ore from eastern Tennessee. Further upriver the Chattanooga bauxite deposits are worked by the giant aluminium refinery at Alcoa. Vast quantities of electric power are required for this enterprise, and for the Oak Ridge atomic energy plant near Knoxville. Knoxville itself is a light manufacturing centre, producing some zinc, textiles, furniture and consumer goods. Power in large quantities is also required for paper mills, the Bowater plant at Calhoun using power from the Hiwassee dam, for example.

Of great significance for agriculture is the production of fertilisers at the TVA National Fertiliser Development Centre at Muscle Shoals. In the 1930's the output was primarily phosphate of lime, produced in limited quantities for a limited market. Over the next thirty years the production of fertilisers expanded enormously, as have the types and forms now produced. The most important of the group are phosphates based on local deposits which are processed into both liquid and granulated forms. Currently some 2,000 farms throughout the nation are demonstrating TVA fertilisers, an indication of the enormous interest that has lifted the use of fertilisers in the United States by over 400 per cent since the Second World War.

4. The Impact of the Plan

Statistics tell of the spectacular changes the TVA has wrought in the Valley since 1933. Although the rest of the nation has also shown progress, the worth of the TVA is reflected in the greater than average increases in incomes and production figures. Very difficult to measure, though readily appraised, are the changes in social welfare. Much of the improvement the TVA claims as a direct result of its co-ordinated planning is reflected in the living standards of the valley people.

i. Farm Size

In such a large area as the Tennessee valley, farm size varies considerably, depending on natural factors of soil and slope as much as the social heritage of the plantation-estate-slave era. There is widespread acceptance of the pre-1933 average farm size of 70 acres as being too small, especially in the limits it placed on desirable routines of mixed farming and rotational planning, for example. The farm size has shown a steady progression upwards, averaging 77 acres in 1949, 98 acres in 1959 and 105 acres in 1966. In some farming quarters there is a trend towards smaller farms as being indicative of more intensive land use, and decreasing size is the proof of greater efficiency. This holds true only when the farm size has been adequate or more than necessarily large in the first place. In the TVA lands the low income levels of 1933—just over US$150 per head per annum—indicated the poverty of the area. In part this was due to inadequate living areas. The present income levels give greater scope for flexibility in the patterns of crops and animals on the farms and consequently a greater degree of safety because of diversification of land use. The early monoculture of cotton, corn or tobacco proved to be highly dangerous economically as well as in a soil usage sense when falling prices coincided with depressed yields in the late 1920's.

Productivity of the farms has also risen as larger areas become incorporated into farming units, even though the numbers of people employed in agriculture have decreased also. Much of the reason for increased productivity is due to the availability of fertilisers and power.

ii. Fertiliser Production

The increase in use of fertilisers was a direct result of the TVA control of the Muscle Shoals plant, the availability of large quantities of hydro-electric power and the education of agriculturalists by TVA farm demonstration programmes. Raw materials for fertiliser production are found north-east of Muscle Shoals on the edge of the Cumberland Plateau.

In 1935 the first significant production of fertiliser began at the Muscle Shoals plant—this was phosphate of lime, a relatively simple phosphate. Over the years the research carried out has produced many more fertilisers and more complex and sophisticated methods for employing them for greater productivity; they now take the form of either powdered, liquid or granulated material of varying sizes. Where necessary the phosphates are enriched with nutrients in which certain soils may be deficient. In the last decade superphosphoric acid has been one of the major developments in the search for greater efficiency in artificial manures.

In 1952, co-operatives and private enterprise concerns were asked to assist in promulgating the latest developments in fertiliser techniques in 36 States, co-jointly with Agricultural Colleges. In 1954 the TVA completed its research into tricalcium phosphate and disposed of its research plant, mainly because sufficient quantities of enriched superphosphate were being produced by other plants. One aspect of this reduction in research into concentrated superphosphates by the TVA was the freeing of research facilities that could be employed elsewhere. As a result of this the Authority commenced, in 1955, full scale production of a 74 per cent concentrated nitrogen phosphate fertiliser known as diammonium phosphate.

The demonstration of TVA produced fertiliser on 50,000 farms has helped transform agriculture in the Tennessee Valley. Farm demonstrations have shown how to employ the correct type of fertiliser as an integral part of farming technique. The fertiliser programme emanating from Muscle Shoals takes in the whole country—it has national as well as regional ramifications, a further example of the widespread influence of TVA. Test demonstrations of TVA fertilisers are regularly being carried out on upwards of 2,000 farms across the United States.

iii. Electric Power

The increase in power consumption in the TVA has been spectacular as a result of the utilisation of hydro sites at first, and later the widespread production of electricity from conventional thermal units. The 1966 consumption figure was an increase of over 50 times the 1933 consumption figure. Of the 1966 production of 84 billion kilowatt hours only 15 billion kilowatt hours were produced from hydro sources, however. The increasing demand for electric power has been so great as to create the necessity for thermal plants and the 80 per cent output they now produce. Cost of power has progressively been lowered so that the consumer price, per unit, is now only about one-fifth of the 1933 price.

iv. River Navigation and Flood Mitigation

In 1966 the Tennessee river carried 2,584 million ton miles of freight, a great deal of it consisting of raw materials up river and finished goods to all parts of the valley. Obstacles which prevented widespread use of the river have been progressively overcome as the main river dams and locks have been constructed. The TVA provides an integral part of the great inland waterway of the Ohio-Mississippi system which serves twenty of the United States.

The value of TVA flood mitigation has been variously estimated, for example, at more than US$100 million worth of savings following exceptional rains in 1963 along the Chattanooga sector of the river alone. Torrents of water spilling out of the Upper Tennessee valley were caught by TVA storage facilities and eased methodically down the river system into the Ohio river. As this was only one of six major floods experienced in the previous 25 years, the real worth of the scheme is obvious. Not only in the Tennessee river, but in the Ohio and Mississippi rivers as well, when floods occur, the huge Kentucky Dam is used to hold back the water from the Tennessee. As a result, flood crests along the Ohio-Mississippi are reduced, preventing great damage especially in the lower courses of the Mississippi. Millions of acres of land behind the Mississippi levees are more productive because they are safe from flood devastation.

It is impossible to estimate the financial savings which have accrued since 1933 because of the TVA flood control system. Neither is it possible to calculate the savings in heartbreak and despair of families which would have been uprooted and scattered by uncontrolled flooding through the length and breadth of the Valley lands.

v. Agriculture

Many aspects of agriculture have been transformed in the Tennessee Valley as a result of the Authority's work. The rehabilitation of agricultural land has been the

Like other TVA dams, the Watts Bar dam, completed in 1942, is multi-purpose in nature, serving the interests of flood regulation, navigation and power production. The powerhouse on the left of the picture has a capacity of 150,000 kilowatts. To the right is the navigation lock and the spillway to cope with flood conditions. The dam is bridged by a roadway for its 2,960 feet length. The 112 feet high dam impounds a lake 72 miles long and 38,600 acres in area.

major transformation, dependent on afforestation of steep slopes and restriction of arable farming to slopes less prone to erosion. Widespread contour ploughing and strip farming techniques have been employed to stabilise grasses. With increased pasture sowing there has been a move away from cropping to mixed farming. Beef cattle are raised on valley farms and fattened on pastures and maize. Maize is still the major grain, together with increasing quantities of grain sorghums. Pastures incorporate clovers, rye grasses and the Asian legumes kudzu and lespedeza, both combining high fodder values with quick growth habits. Dairying is widespread around the major centres of population, and is often combined with pig raising, poultry production and some cash cropping.

The valley is still a major producer of the other cash mainstay crops—cotton and tobacco. Greater rotation of these crops is now practised, however, and various combinations include potatoes, peanuts, soybeans and smaller acreages of vegetables. Pasture and hay crops play a major role in these rotation patterns, whereas previously they were almost absent. Poultry production has been a major innovation in the valley—broiler growing in particular. Orcharding, of stone fruits and pome and berry fruits, is a feature of the slopes. Apples, pears and the berries generally occupy the higher slopes. Some sheep farming is carried out where higher slopes have not been fully planted to trees, although in the main the outstanding feature of the slopes has been the pressure on taking such land out of production, and millions of acres are now planted to a variety of trees.

vii. Some Final Considerations

Today the Tennessee Valley Authority's water, power, reafforestation, agricultural and industrial development has transformed the lives of the people of the valley. Soil erosion has been arrested, catchment areas covered with vegetation, farm production greatly increased through improved methods and the use of fertilisers, electricity has been supplied to nearly all farms, floods largely controlled and a great inland waterway constructed.

The TVA's work continues with Tributary Area Development to assist local communities. Other projects include the development of the land between the lakes in western Kentucky and Tennessee. The TVA has been assigned 170,000 acres here to develop a nation-wide demonstration area showing the maximum utilisation of natural resources for the furtherance of conservation education.

It was against the background of future development and the widespread impact of TVA planning that President John F. Kennedy said on 18 May 1963, 30 years after the Authority was inaugurated:

In short, the work of TVA *will never be over. There will always be new frontiers for it to conquer. For in the minds of men the world over, the initials* TVA *stand for progress . . .*

Bibliography

Davis, D. H., *The Earth and Man*, Macmillan Company, New York 1955.
Pickles, T., *The Wealth of the World*, J. M. Dent & Sons, London 1957.
Pounds, N. J. G., *Political Geography*, McGraw Hill, International Students Edition, Kogakusha Co. Ltd, Japan 1963.
Smith, J. R., and Phillips, M. O., *North America*, Harcourt, Brace and Company, New York 1942.
Suggate, L. S., *World Geography Human*, Thomas Nelson and Sons Ltd., London and Edinburgh 1958.
White, C. L., and Foscue, E. J., *Regional Geography of Anglo-America*, Constable, London 1956.
Wright, A. J., *United States and Canada*, Appleton-Century-Crofts, New York 1956.

Questions

1. 'The main task of the TVA was the improvement of navigational facilities and the control of flooding. The scope covered by the Authority turned out to be much greater than this.' Critically examine this statement.
2. Why has the TVA become world famous ? What lessons has it to teach people living in backward areas in other countries ?
3. What were some of the problems facing the TVA in its efforts to rehabilitate the Tennessee River valley ? Explain how these obstacles were overcome.
4. Explain fully the work done towards controlling erosion and run-off. Why were these factors so important to the Tennessee Valley scheme ?
5. 'The impact of the plan has been far reaching, beyond the boundaries of the Tennessee Valley itself.' Explain fully the meaning of this statement and cite specific examples to illustrate your answer.
6. How has the Tennessee Valley project demonstrated administrative direction and the resultant effects on land use patterns ? What are the main changes which have been initiated in the practice of agriculture by the Tennessee Valley scheme ?
7. 'In the Tennessee Valley farm size has been slowly rising, but not entirely as some farms have become smaller.' Explain fully the reasons for these different trends. How does this situation compare with farming areas in Australia ?